Plant Monitoring and Maintenance Routines

Based on papers presented at the one-day Seminar *Plant Monitoring and Maintenance Routines*, held at the IMechE Headquarters, London, UK, on 24 October 1996.

IMechE Seminar Publication

Plant Monitoring and Maintenance Routines

Organized by
The Tribology Group of
The Institution of Mechanical Engineers

IMechE Seminar Publication 1998–2

Published by Professional Engineering Publishing Limited for
The Institution of Mechanical Engineers, Bury St Edmunds and London, UK.

First Published 1998

ISSN 1357–9193
ISBN 1 86058 087 4

A CIP catalogue record for this book is available from the British Library.

Printed by Antony Rowe Limited, Chippenham, Wilts, UK.

Contents

Related Titles of Interest

Title	Author	ISBN
Process Machinery – Safety and Reliability	W Wong	1 86058 046 7
An Introductory Guide to Industrial Flow	R Baker	0 85298 983 0
An Introductory Guide to Pumps and Pumping Systems	R K Turton	0 85298 882 6
The Reliability of Mechanical Systems	J Davidson	0 85298 881 8
Fluid Sealing	Edited by B D Halligan (BHR Group Publication 26)	1 86058 091 2

For the full range of titles published by MEP contact:

Sales Department
Professional Engineering Publishing Limited
Northgate Avenue
Bury St Edmunds
Suffolk
IP32 6BW
UK

Tel: 01284 763277
Fax: 01284 704006

S472/001/96

Maintainability requirements and standards for mechanical engineers

R H UNWIN BSc, CEng, MIMechE
Andover and BSI Chiswick, UK

INTRODUCTION

This paper gives a brief explanation of how IEC became involved with R&M standards. This leads on to the IEC 706 series of Guides on Maintainability and Maintenance Support.

Work is currently in hand to update the IEC range of standards on AR&M, or Dependability, which is the official IEC terminology? This work is intended to make the whole subject more cohesive and is linked with the work of other organisations including ISO and the European CEN and CENELEC organisations.

IEC

IEC, the International Electrotechnical Commission, was founded in 1906, and is a truly international organisation with its headquarters in Geneva with representation from most of the industrialised countries of the world.

The National Standards organisations, which in UK is BSI, provide the focal point in each country, nominating members to represent their country at Technical Committee meetings and on Working Groups. They also accept draft documents from IEC, arrange for them to be circulated for comment and collate the comments from their country.

TC 56

Technical Committee 56, with the title of "Reliability of Electronic Components and Equipment" was established in 1965. With the development of reliability prediction techniques, which largely grew out of the electronics industry, considerable expertise was being accumulated in the field of reliability and TC 56 was therefore formed with the specific task of developing reliability standards.

Every 12 to 15 months a Plenary meeting of the whole committee is held where all the work is discussed and progress noted. New work is considered and time is allowed for the Working Groups to meet.

WORKING GROUPS

As the range of work undertaken by TC56 expanded, Working Groups were initiated to develop the appropriate standards. The WGs are comprised of members from different countries with a Convenor, or Secretary, responsible for the work of the group. Most groups have 10 to 20 members, but some members are more active than others. WGs usually have one or more meetings between the main meetings, but much work is done by correspondence and increasingly by e-mail.

The current list of Working Groups is:

WG1	Terms and Definitions	
WG2	Data Collection	- Disbanded
WG3	Equipment Reliability Verification	
WG4	Verification and Evaluation Methods	
WG5	Formal Design Review	- Disbanded
WG6	Maintainability	
WG7	Component Reliability	
WG8	Reliability and Maintainability Management	
WG9	Analysis Techniques for System Reliability	- Disbanded
WG10	Software Aspects	
WG11	Human Aspects of Reliability	
WG12	Risk Analysis	
WG13	Project Risk Management	

This list shows the range of work that TC56 has now taken on. The ones that have been disbanded are generally the ones which did the original reliability work, and that is mostly finished. The new standards tend now to be more in the management field.

Working Group 6

WG 6 on Maintainability was formed in the early 1970s and was originally set up to produce a comprehensive document on Maintainability, which became known as IEC 706.

Structure of IEC 706

The various sections were developed separately but the final composition of IEC 706, with the 9 sections issued in 6 Parts, was:

Part 1, (IEC 706-1), issued in 1982, containing the following sections:

Section 1 Introduction to Maintainability

Section 2 Maintainability Requirements in Specifications and Contracts

Section 3 Maintainability Programme

Part 2, (IEC 706-2), issued in 1990, containing the following sections:

Section 5 Maintainability Design Studies

Part 3, (IEC 706-3), issued in 1987, containing the following sections:

Section 6 Maintainability Verification

Section 7 Collection Analysis and presentation of data related to maintainability

Part 4, (IEC 706-4), issued in 1992, containing the following sections:

Section 8 Maintenance and Maintenance Support Planning

Part 5, (IEC 706-5), issued in 1994, containing the following sections:

Section 4 Diagnostic Testing

Part 6, (IEC 706-6), issued in 1994, containing the following sections:

Section 9 Statistical Methods in Maintainability Evaluation

I should make the point at this time that these documents are Guides, not Standards which have to be followed, as for example technical standards on electrical sockets. The wording of the guides is therefore strictly controlled by IEC and we must not use the words "shall" or "are to", but "should", as they are advisory documents.

DESCRIPTION OF IEC 706 SERIES

I will spend a few minutes talking about the contents of each of these, although some are of more interest than others.

Part 1 Introduction, requirements and maintainability programme

Part 1 is a comparatively thin booklet, but contains the first three sections.

Section 1 is a brief introduction which explains the concept of maintainability and how it is applied through the development cycle.

Section 2 covers the maintainability requirements in specifications and discusses the scope of what has to be considered when maintainability characteristics are included as requirements in the development or acquisition of an item.

Section 3 describes the content of a Maintainability programme and was written for the case where a contractual relationship exists between the contractor and the customer with a stated maintainability requirement.

This part therefore covers the introduction, how a maintainability requirement should be written and how a programme can be set up to meet that requirement.

Part 2 Maintainability studies during the design phase
Part 2, Maintainability Studies during the Design Phase, took a long time to be accepted and was issued three years after Part 3.

This outlines the maintainability studies to be performed during the preliminary and detailed design phases of a project and forms part of the maintainability programme described in Section 3 Part 1.

The objectives of these maintainability studies are to predict the quantitative maintainability characteristics of equipment, to identify changes required to the design to improve maintainability and to guide design decisions. It also provides guidance for design support activities and maintainability checklists for design criteria and design reviews.

Part 3 Maintainability verification and analysis and presentation of data
Part 3 contains Sections 6 and 7.

> Section 6 describes the various aspects of verification necessary to ensure that the specified maintainability requirements have been met and provides suitable procedures and test methods.
> Section 7 considers the collection, analysis and presentation of maintainability related data. Such data is typically collected by the supplier during development and much of that may be required to support the verification process.

This part therefore looks at how closely the requirement has been satisfied. These activities will take place during development with contractors trials and may go on to a maintainability demonstration to assure the customer that his requirements have been met.

Key maintainability data of concern would typically be corrective and preventive active maintenance times with clear definitions of the personnel and facilities required.

Part 4 Maintenance and maintenance support planning
Part 4 on maintenance and maintenance support planning was a late addition to the 706 series, but is an important element. It marks a change of policy in IEC, with a move away from analytical studies towards more practical applications of project management.

Part 4 was based on the US Mil Spec 1388 and describes the tasks for planning of maintenance and maintenance support which should be performed during the design phase of a project in order that availability objectives in the operational phase can be met.

Efforts to reduce the active maintenance time by maintainability programme activities described in Part 1 and the associated studies in Part 2 need to be accompanied by the reduction of non-active maintenance time elements caused by technical, logistic and administrative delays. There is therefore a need to develop a maintenance concept, perform maintenance planning analyses and determine the maintenance support resources required, such as personnel and training, test and support equipment, spare parts and technical manuals.

It includes a useful Annex on the determination of maintenance support resources.

Part 5 Diagnostic testing
Part 5 on Diagnostic Testing provides guidance on the early consideration of testability requirements and shows how these can be integrated into the operation and maintenance of a system.

It was intended that this section would be applicable in concept and principles to all categories of equipment and not just electronic. Although there is more scope for BITE on electronic equipment, most modern mechanical systems have sophisticated electronic measuring and monitoring devices which allow the use of much improved diagnostic testing facilities.

Part 6 Mathematical procedures

Part 6 contains the mathematical procedures and expressions used in Part 3, Section 6 on verification and in Part 5 on diagnostic testing. It is now a fairly slim document, but in earlier days the mathematical content was very much greater. Not a lot of that now survives in the document as many of the mathematical methods can be found in textbooks and therefore were not considered appropriate to a standard.

MAINTAINABILITY REQUIREMENTS

Having set the background and told you of the existing standards I would now like to return to maintainability requirements. Although, as you will see from the breadth of the 706 series standards, this is only the first step in the development of a product from the maintainability point of view, it is an important one. Information on maintainability requirements is given in IEC 706-1, Section 2.

First, what is maintainability? The definition is "the probability that <u>a given maintenance action</u>, for an item <u>under given conditions of use</u> can be <u>carried out within a stated time interval</u>, when the maintenance is performed <u>under stated conditions</u> and using <u>stated procedures and resources</u>."

So you have there the basis of a requirement:

For a given maintenance action:
- under given conditions of use
- carried out within a stated time
- performed under stated conditions
- and using stated procedures and resources

The last four items must be stated and therefore included in the requirement.

There are a number of levels to the requirement, ranging from the broad maintenance policy to the detailed maintainability specification.

Maintenance policy

First the client's maintenance policy. The client will need to describe his general maintenance policy by giving information on how his support organisation works, such as:

- a definition of his levels of maintenance.
- the depths of corrective maintenance envisaged at each level.
- the environmental conditions at each level.
- the resources available or planned at each level.
- limitations of preventive maintenance.
- maximum acceptable turn-around times.

- training of maintenance personnel and of operators, where they are to be involved in maintenance tasks.

Another important consideration these days is the relevant legal or statutory requirements including health and safety.

Maintenance concept

Next the maintenance concept. The client should have stated his broad requirements for the support of the specific product in the form of a maintenance concept, including:

- the repair plan;
- the planned life;
- the quality of the product and the standard of maintenance required;
- the requirements for technical maintenance manuals;
- the ability for the product to be upgraded or modified;
- the method of disposal at the end of its life.

The maintenance policy and concept documents then become the basis for the preparation of a Maintainability Specification.

Maintainability specification

The maintainability specification gives the detailed maintainability requirements for the product, and provides the basis of understanding between the customer and the supplier, which may be incorporated into the contract.

The maintainability specification will be divided into a number of parts, depending on the complexity of the product:

- the quality maintainability requirements (explanation of product functions, fault definitions, environmental and operational conditions, planned life); i.e. the conditions.
- the repair organisation (levels and depth of repair, personnel, skills) i.e. procedures and resources.
- the quantitative measures of maintainability performance. The maintainability is usually defined in terms of Mean Active Repair Time or MART, with a maximum time also given. i.e. stated times.

It may also have:

- detailed requirements, including test procedures, access, use of special tools and test equipment.
- relationship with any existing maintenance support.

How this specification will be met is the responsibility of the designer. He should implement a maintainability programme, (IEC 706-1, Section 3), using the study techniques in IEC 706-2 and ensure that his design is easy to maintain with the necessary maintenance support, IEC 706-4 and that it can be repaired within the required times. The maintainability will then be verified in accordance with IEC 706-3.

FUTURE STANDARDS

I would like to spend the rest of the time telling you briefly about what standards are in development and where we are going.

RCM

The first is a Guide on Reliability Centred Maintenance or RCM which started life in about 1989. The methodology described in this guide is based largely on the document "MSG-3" which was produced by agroup of engineers from the US airlines, together with the manufacturers, to rationalise the preventive maintenance on the more complex aircraft, then entering service. MSG-3 is a descendent of MSG-1, the original document on RCM published in 1969.

The IEC RCM Guide addresses the preparation of an initial preventive maintenance programme for new systems or structures, but it can also be applied to existing ones.

This document had a bit of a rough ride through the early draft stages as it had some limitations and there are a number of procedures in use in different industries, and although all were based originally on the MSG document, variations have been introduced and there are now a number of experts around who expect to see it done their way. However we have extensively re-written it and it is about to be re-issued as a draft for international comment.

Guide to ILS

The other new work we have in hand is the Guide to Integrated Logistic Support (ILS). This was proposed as a commercial version of US Mil Spec 1388, which would be applicable to non-military projects and would show the value of integrating logistic support into the design process. The perception of the existing military ILS procedures, probably rightly as they are now applied, is that it is expensive and can only be considered for military projects.

We believe that there is scope for a guide on this subject, as if sensibly applied the concept of ILS is useful and ensures that the logistic support is properly considered. Where it usually goes wrong is that it is not tailored strictly to the requirements of the project and much more paper is generated than necessary.

The first draft was not well received on the limited circulation it had as there was too much military terminology, and from the comments it was clear that readers did not really understand the objective. It has now been rewritten and we reviewed it at the last WG meeting and believe that it is much improved.

RESTRUCTURING OF IEC STANDARDS

In the late 1980s a decision was taken to change progressively the structure and the numbering system of IEC publications.

Previously each standard took the next available IEC publication number when it was published, with the result that Dependability Standards were scattered throughout the IEC catalogue and the relationship between these various standards was difficult to recognise. A new approach to numbering, called the "Toolbox Concept" was therefore introduced as shown in the diagram.

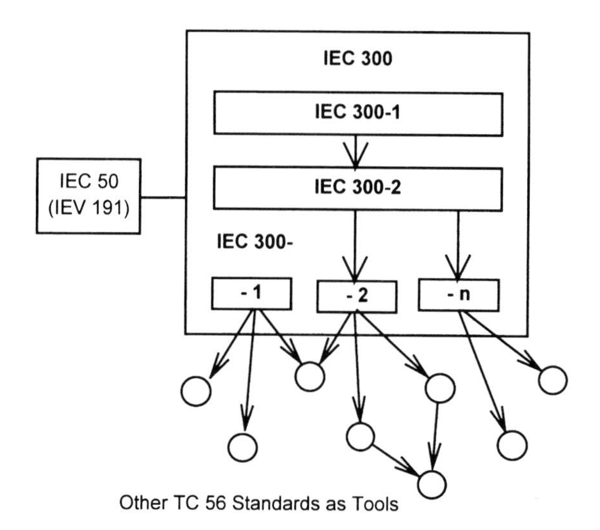

TC56 Toolbox Concept

The toolbox concept implies a four level arrangement of TC56 standards with IEC-300 on A,R&M Management used as the focal point, equivalent to the use of ISO 9000 which is the focal point for standards on Quality management.

The IEC 300 series produced by TC56 describe typical Dependability Programmes.

IEC 300-1 - Dependability Management - Dependability Assurance for Products This is intended for contractual purposes and is dual numbered as ISO 9001-4. It is the top level document which covers the requirements on dependability assurance and programme elements within the supplier's organisation, with links to the customer's organisation.

IEC 300-2 - Dependability Management - Dependability Programme Elements and Tasks This has been published this year and describes in more detail the establishment of a dependability programme.

IEC 300-3 is a series of "Application Guide" Standards. An Application Guide will aid the user, that is the user of the standard, in the choice and application of proper tools for a particular situation.

Tools are freely numbered, self-contained documents and many of the existing IEC documents on specific procedures and methodologies come into this category.

MAINTAINABILITY APPLICATION GUIDE

As a result of the new standards structure, IEC 706, the Maintainability Standard, found itself without a niche in the hierarchy of standards. A Maintainability Application guide was therefore required at Level 3 with some maintainability tools at the next level.

It was decided that a new Maintainability Guide could be based on IEC 300-2. Some guidance on maintainability was required, as 300-2 referred almost exclusively to "Dependability" and was not specific about the elements which make up dependability, ie reliability and maintainability,

The Maintainability Application Guide was therefore written under the headings of IEC 300-2, and dealt with the maintainability aspects of the various tasks which make up the dependability programme. It is perhaps not quite the right layout for a treatise on maintainability, but it fits the pattern of the 300 series documents and there are references throughout to the sections of IEC 706 to be used as tools on specific topics.

The Maintainability Application Guide, now numbered IEC 300-3-10, has been issued as a Committee draft on its first circulation and comments have been received back. It seems to have been reasonably well received and we incorporated the comments at the recent Working Group meeting.

We hope that this will be accepted for publication late in 1997 and will be the new IEC Guide on Maintainability, supported by the existing 706 series.

CEN MAINTENANCE STANDARDS

CEN Committee 319 are in the process of writing some new Standards on Maintenance, as opposed to Maintainability which is the province of WG 6. They have set up 5 Working Groups and the work is aimed particularly at contracted maintenance services, defining the task, the contract and its supervision. They are currently in draft form and cover the topics shown:

Classification of Maintenance Services
Contractual Maintenance
Terminology
Quality Assurance Requirements for Maintenance Operations
Documentation for Maintenance

SUMMARY

In summary, the maintainability standards which are available to industry are contained in IEC 706, which BSI sells as a dual numbered standard BS 6548. In addition, IEC 300-1 and IEC 300-2 are now available which give the basis of a dependability programme which can be used in the development of a product.

We are developing the new Maintainability Application Guide, Reliability Centred Maintenance and Integrated Logistic Support, which we hope will form a structured series of standards which will be of constructive help to industry.

ABOUT THE AUTHOR

R H Unwin left the Army about 2 years ago, where he served for 30 years in REME, the Army's maintenance corps, as a mechanical engineer. Since then he has been working as an independent engineering consultant in the field of Logistic Support in the defence industry. He became involved in the Standards business while in the Army in 1986 by being appointed as a member of a BSI Committee which dealt with AR&M standards.

Since that time he has been a UK representative on the IEC (International Electrotechnical Committee,) Technical Committee 56 which produces standards on Reliability and Maintainability. For the last 4 years he has been the Convenor of the TC56 Working Group 6, developing standards on Maintainability.

S472/002/96

Condition monitoring – the mixed economy

B G HUDSON and **A WASSON**
ICI Chemicals and Polymers Limited, UK

1 INTRODUCTION

1.1 ICI on Teesside

ICI on Teesside consists of three geographically remote sites, separated by the River Tees with a maximum travelling distance of 12 miles between sites. There are over 30 production units, ranging from large single stream, continuous process plants to small, batch units, producing a range of petrochemical, plastic polymers, performance chemicals and fertilisers, either for external sales or feedstocks for other production facilities within ICI.

There are over 12,000 rotating and reciprocating machines in use on Teesside ranging from 50 MW compressor trains to conventional simple centrifugal process pumps.

1.2 The history of machinery condition monitoring within ICI on Teesside

ICI has always made use of condition monitoring techniques as new technology became available. Currently, the majority of available techniques; such as vibration, oil analysis, temperature (direct measurements and thermography), fixed monitoring systems, non-destructive technologies (ultrasonic crack detection and thickness measurements), log sheet trending, efficiency surveys, acoustic emission and motor current waveform analysis; are in use somewhere in ICI on Teesside. Historically, the application of the techniques would generally be in response to particular problems and provided by specialists from the central engineering organisations, with very little done on a routine basis as predictive technologies. Lubrication oil analysis was established as a contract routine service in 1990. Vibration monitoring was established as a routine service in 1993, based on frequency spectrum analysis.

2 THE ESTABLISHMENT OF A CENTRALISED VIBRATION MONITORING SERVICE

2.1 The basis of the service

Vibration monitoring and lubrication oil analysis are the most universally applicable of predictive technologies to the types of machines used within ICI Teesside. By 1993, oil analysis was already an established service with fairly comprehensive coverage. The establishment of the comprehensive use of vibration monitoring as a predictive technology was the next logical step. Equipment and analysis software to do frequency spectrum analysis on a routine basis had been purchased in the mid 1980s. The need to provide trained and experienced specialist staff plus the availability of existing equipment dictated the establishment of a centralised service.

Selection of machines for routine vibration monitoring was based on the potential impact of a failure, in terms of safety and environmental issues, production loss and machine repair costs. Initiatives such as Critical Machines (1) and Machinery Care (2) catalysed this approach.

2.2 The mechanism of development and growth

The initiative to establish a routine vibration monitoring service came from the same section that provided the technical and practical support service to the plants on ICI Teesside for machines, thus facilitating initial discussions with the customer plants.

Potential customers were categorised in terms of existing interest and potential for generating benefits. A programme, spread over three years starting in 1994, was developed to establish each customer within the service. Initial meetings were organised against this plan with each potential customer at which a presentation on the techniques and benefits of condition monitoring was made. Having generated interest, a list of machines that would benefit from routine vibration monitoring was generated and routines implemented.

Key features within this approach were:-

- The development of a customer and supplier relationship.
- Jointly identifying the customer's condition monitoring requirements.
- Formal service level agreements.
- Implementation and operation imposing a minimum impact on the customer's local resources.
- Provision of the resources to do routines surveys routinely.
- Provision of a rapid response to requests for non-routine surveys.
- Actively seeking feedback on successes and failures.
- Evaluation of benefits, costs and provision of feedback.
- Regularly reviewing the effectiveness of the service with the customer.
- Application of standards and procedures from the start (ISO 9002).

Targeting of the service to each machine's needs and potential for payback generally has resulted in about 10% of each customer's machines being covered by the service.

Figure 1 shows the growth achieved to date. Figure 2 shows the benefits generated to date.

2.3 Why centralisation?

Centralisation allowed the rapid growth of vibration monitoring across ICI's Teesside plants by optimising the use of expensive equipment and scarce skilled resources without compromising standards. Central management allows the consistent management of safety and the application of standards. Dissemination of learning is simplified and the retention of corporate knowledge is ensured. Costs can be minimised, particularly by taking advantage of the economies of scale. Technology improvements can be more easily justified on a macro scale. The resources of a centralised service suffer less from conflicting priorities, thus ensuring that routine surveys happen on a routine basis.

2.4 Why contractors?

Carefully selected contractors come with an initial level of experience and expertise, thus minimising initial training requirements. Contractor resource levels can be more easily managed to suit variable demands. The combination of these two factors facilitates rapid growth. Using a number of contractors creates a competitive costing structure.

2.5 Why ISO 9002 registration?

The discipline of ISO 9002 Registration ensures that the service to each customer is defined and the customers requirements are met. It helps to ensure a consistent set of standards across the service and facilitates rapid growth without compromising these standards. It facilitates the implementation of change with control. Finally, it imposes an internal and external auditing structure.

2.6 Sustaining the service

Regular customer reviews ensure that the customer's requirements always reflect current needs. By identifying the benefits received by the customer, interest in maintaining and improving the service and the customer's gains from condition monitoring can be maintained. It is necessary to be continually searching for ways of improving the service and reducing costs by taking best advantage of predictive technology techniques and developments in Information Technology.

3 THE CONCEPT OF THE MIXED ECONOMY

3.1 Levels of sophistication

Vibration monitoring techniques can be ranked in terms of sophistication. In general terms, the more sophisticated the technique, the better the identification of the fault and the earlier it can be detected.

Low Sophistication

Overall vibration measurements often linked with bearing “health” measurements.

Medium Sophistication
Frequency spectrum analysis and enveloped frequency spectrum analysis, identifying specific faults.

High Sophistication
Techniques such as Real Time Measurements, Reciprocating Compressor Indicator Diagrams and Modal Shape Analysis.

Each technique requires different levels of training and experience for the operatives. Increases in sophistication result in higher survey costs, due to higher labour charges.

Higher levels of technique sophistication should not automatically be rejected purely on a cost basis. These techniques also generate greater levels and sophistication of information. This allows faults to be detected or predictive information to be gained that simple techniques will not provide. High sophistication might be the only way of gathering the necessary information. The level and precision of the information gained from the higher sophistication techniques may allow greater intervals between surveys, thus giving a lower overall survey cost. The benefits gained should still exceed the costs.

The choice of technique depends on evaluating the cost of failure versus the benefits of early detection and detection success rate. Low sophistication techniques have the potential benefit of quantifying and enhancing "Look, Listen and Feel" inspections by operating staff.

Continual review, based on "Mean Time Between Failure", "Root Cause Analysis" and "Impact of Failure", ensures that the correct level of technique is being applied to a given machine. A given machine could be subject initially to "Low Sophistication" techniques, to "Low" and "High Sophistication" techniques during a period of problems or investigation and revert to a "Low Sophistication" technique when the problems are understood. Similarly, the frequency of surveys could vary between techniques and over the life of the machine.

In ICI "Medium Sophistication" techniques generally only provide a payback on about 10% of the machine population. The concept allows a higher percentage of machines to be subject to some level of monitoring without incurring unnecessary costs.

Figure 3 graphically illustrates the principles of the Mixed Economy.

4 AN APPLICATION OF THE MIXED ECONOMY

4.1 A typical production area

One of the Production Areas at Billingham consists of six plants. Four of the plants are single stream, continuous units whilst the other two are batch units. There is a total population of approximately 250 machines, 90% of which are centrifugal pumps.

The largest plant within the production area is a continuous, single stream unit that supplies many of the other plants within the area. The reliability of the plant is crucial as downtime can be cascaded on to the other plants that it supplies.

This plant has approximately 60 machines, many of which are 'canned' units or high speed Sundyne pumps. Historically these machines had suffered severe reliability problems, to such an extent that it was not uncommon to change some units several times a week.

These reliability problems, allied to the need for high levels of plant availability, made this plant an obvious choice for machinery condition monitoring. Regular spectrum analysis has been in place since 1994, with the 'mixed economy' approach being adopted in 1995.

4.2 THE MIXED ECONOMY IN ACTION

4.2.1 Frequency spectrum analysis

Frequency Spectrum Analysis is well established in the Production Area. All the plants are covered to some extent, with the service being provided by the Teesside Machines Condition Monitoring Service.

Initially there was no well defined criteria for selecting machines for analysis, it was simply a qualitative assessment based on:

1. The perceived reliability of each machine.

2. The potential impact of a failure on plant availability.

The machines originally selected for analysis were monitored every two months. Other problematic machines were surveyed on an 'ad-hoc' basis to gather information to support failure investigations.

4.2.2 Spectrum frequency analysis - refining the approach

Although the original Spectrum Frequency Analysis routines were providing some useful data, there were a number of shortcomings:-

1. The two monthly survey was too infrequent to predict some failures.

2. The process for selecting machines for analysis was not rigorous, meaning that effort was not well focused.

3. The technique required skilled operators/data interpretation making it relatively expensive to apply on a widespread basis.

In the light of the above problems the system was revised in the following ways:-

1. The process for selecting machines was revised to include the following facets:-
    a) Collection of mean time between failure data to identify those machines which failed most often and assign appropriate survey frequencies.
    b) Production loss accounting techniques were used to identify those machines whose failure had a significant impact upon production.

c) Data from 'Critical Machines Assessments' (2) was used to highlight machines where detailed monitoring could have a safety benefit.
d) A consideration of the cost of failure versus the benefits of early detection.

2. The introduction of overall vibration level monitoring using simple data collectors.

4.2.3 Overall vibration level monitoring & bearing health checks
Routines have been established to take overall vibration level readings and bearing shock pulse measurements. The measurements are taken and stored using a hand held data collector, and are subsequently downloaded into a PC for analysis. The data collection and analysis is undertaken by a plant technician. This helps to promote an 'ownership' of the plant machinery and the regular tours can identify other problems at an early stage, for example low oil levels or seal leaks.

For a typical centrifugal pump set, overall vibration readings are taken from the motor and the pump, and shock pulse measurements are taken from all the rolling element bearings. These measurements are plotted against time by the computer software. In this format it is easy to monitor vibration trends towards alarm level, enabling an estimate of the likely failure date to be made. This makes it possible to plan maintenance and remove pumps for repair before they fail. Experience within the Production Area has shown that removing a pump for overhaul before it fails saves 20 - 70% of the complete rebuild cost following a failure. This saving is small when compared with the benefits of improved plant availability and fewer unplanned production losses. Figure 4 shows a typical trend.

4.2.4 Using The Mixed Economy
All the process pumps on the plant are included on a weekly overall vibration monitoring routine. The rapid and low cost nature of the technique makes it feasible to do a large number of machines so frequently. The regular surveys make it possible to develop good trends, and make accurate predictions about future failure dates.

Frequency Spectrum Analysis is used in a more focused manner to support the above routines, viz.:

1. On safety or environmentally critical machines.
2. Machines with frequent, costly breakdowns where the data from spectrum analysis can usefully be Incorporated into a failure analysis assessment.
3. To help identify problem areas to aid the selection of measurement points for overall vibration level readings. For example, spectrum analysis may have shown that a high speed centrifugal pump has a history of gearbox problems, and that the unit fails when the gearbox vibration reaches 12 m/s RMS. This information can be used to ensure that the overall vibration readings are taken from an appropriate point on the gearbox, and that the alarm level is set at 12 m/s RMS. This avoids the need for regular spectrum analysis, but ensures that the overall vibration readings areas focused as they possibly can be.

4.3 MIXED ECONOMY SYSTEM – THE BENEFITS

4.3.1 Widespread condition monitoring

The rapid, relatively low cost nature of overall vibration level trending means that a wide range of machines can be regularly assessed. The routines are carried out by local plant tradesmen which promotes an ownership of the machinery.

4.3.2 Improved reliability and availability

The trending facility of overall vibration level readings makes planned maintenance possible. This reduces the number of unplanned plant outages and improves the availability of the assets. See Figure 5. The data from spectrum analysis makes it possible to ascertain the root cause behind many failures. With this information available it is possible to modify the machines or change operating practices to improve the reliability of individual units.

4.3.3 Lower Costs

Assessing what technique should be applied to each machine means that the relatively sophisticated and costly spectrum analysis method is not applied unnecessarily. Fewer failures, combined with a shift towards a planned and predictive maintenance regime, mean that maintenance costs are reduced. The biggest gain, however, is the improved plant reliability and uptime.

5 FUTURE DEVELOPMENTS

Within the next 12 months we expect the period of high growth to come to an end. We are continually looking for opportunities to update and enhance the capabilities of the service and generate even greater benefits for our customers. Over the next year, we expect to extend the concept of the Mixed Economy to other customers. We are continually evaluating new techniques and introducing those that provide added value to the service. The integration of all predictive technologies into one service, operated and accessed via computer networks, is seen as a key short term objective. Further reductions in operating costs will be sought and methods for evaluating the benefits from holding the gains made to date will be developed.

REFERENCES

1. Machinery Safety – Registration and Verification of Critical Machines, Maximising Rotating Reliability, HB Carrick and JJ Lewis, IMechE HQ, 8 December 1994.

2. Machinery Care – An Operators Approach to Reliability, N H Wells, 6th European Congress on Fluid Machinery, The Hague, Netherlands 13/14 May 1996.

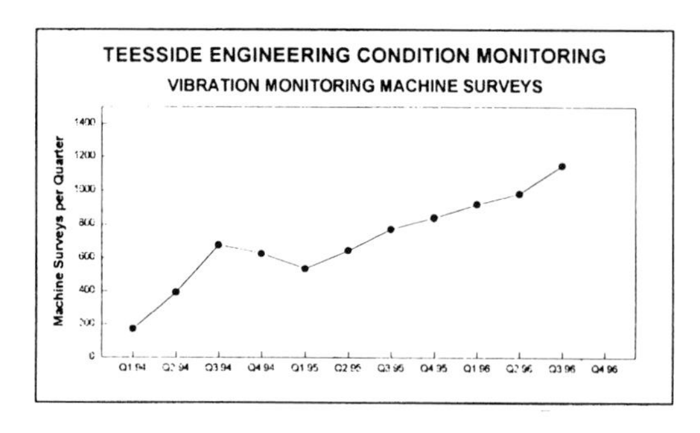

Figure 1. Vibration monitoring service growth

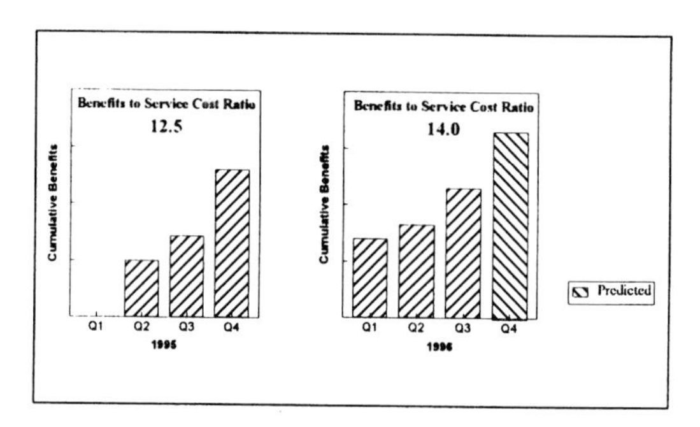

Figure 2. Benefits gained to date

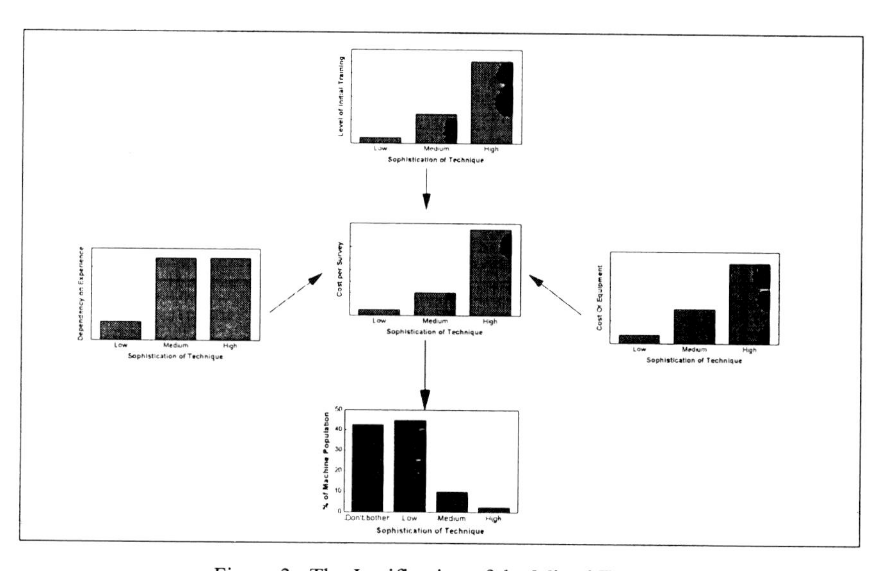

Figure 3. The Justification of the Mixed Economy

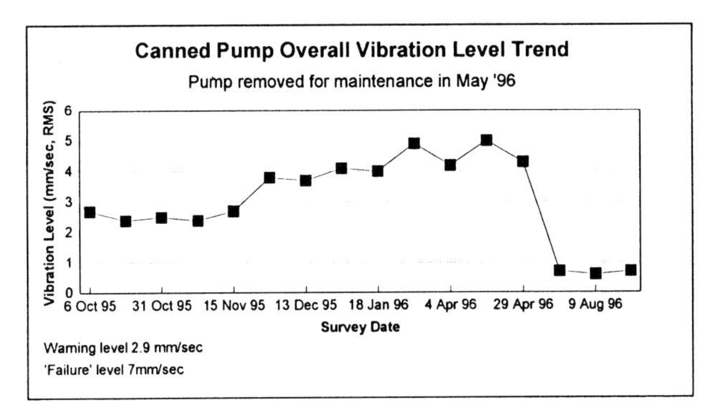

Figure 4. A typical trend

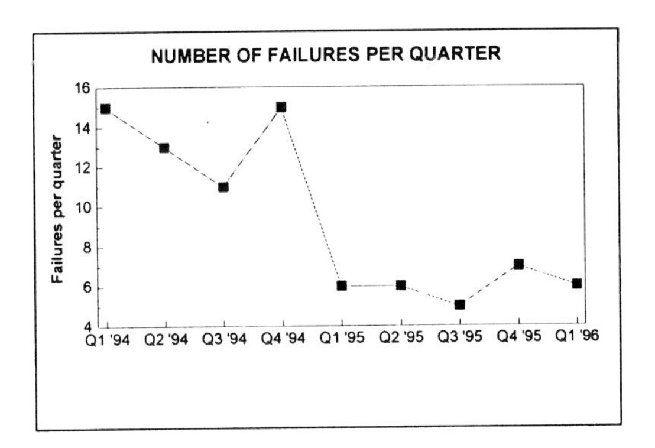

Figure 5. Fewer Failures

S472/003/96

Infrared thermography in maintenance

M BOSWORTH
AGEMA Infrared Systems Limited, Leighton Buzzard, UK

INTRODUCTION

It is often difficult to measure the costs incurred from equipment failure – be it a broken electrical connection in a switch panel or an overheated motor in an automated production line. It is even more difficult to quantify the financial benefits gained by preventing equipment failure. But when more and more companies report huge revenue losses due to unplanned shutdowns, there can be no doubt about the necessity for predictive maintenance.

One maintenance technique which is seeing increasing popularity amongst maintenance engineers is infrared thermography. As a non-contact temperature measurement technique, infrared condition monitoring enables companies to detect a variety of potential faults *before* they have a chance to fail and disrupt production.

FOCAL PLANE ARRAY DETECTORS

The reason for the growing interest in this technique is due largely to the introduction of a new generation of lightweight thermal imaging systems incorporating Focal Plane Array detector technology. With the new benefits provided by these cameras (compared to traditional systems incorporating single detector scanning technology), more and more companies are choosing to adopt thermographic inspection techniques within their plant condition monitoring programme.

SIZE AND WEIGHT BENEFITS

One of the most significant advantages of replacing single detector scanning systems with solid state matrix arrays is the reduction in size and weight which can be achieved in these cameras. Some FPA systems are now so small and light that they can be held and operated in one hand only, enabling operators to proceed through a survey much more quickly than before. Whilst many companies have resisted the use of infrared thermography for condition monitoring purely on the grounds of the size, more and more companies will now start to use the technique once they realise how much more practical they have become to use.

IMPROVED FAULT DETECTION

Those companies which have in the past doubted the effectiveness of infrared techniques for detecting faults will be left in no doubt about the ability of FPA cameras once they see the superior image resolution now provided by these systems. With four times the image resolution of scanning systems, the new FPA cameras are able to detect much smaller faults than before, faults which might otherwise have led to costly and unplanned shutdowns.

The ability to store image data in 12-bit format is becoming a standard feature in new FPA cameras which customers should welcome if they want to be absolutely sure of not missing any faults. Even if an operator suspects there may be a fault but doesn't have time to make the correct camera settings to check it out in the field, 12-bit image storage will able him to store an image of the suspect component whilst he is on site and analyse it in full at a later date. So no matter what the camera settings at the time the image is stored, the image can be recalled and manipulated at a later date to highlight any hot spots previously undetected during the survey.

The use of PC-card hard disks for storing images is now becoming the standard technique for storing image data. Many hundreds of thermal images can now be stored on Type II or Type III PC-cards.

COMMERCIAL PRICE

The use of FPA detector technology has also made it possible for customers to buy imagers which offer the same levels of image quality as military systems but at a much more commercially viable price. The new FPA systems will prove particularly popular with customers looking for high resolution image quality but whose budgets up till now have been limited to paying for much lower resolution products such as pyrovidicon cameras.

EASE OF USE

As Focal Plane Arrays become the established technology amongst thermal imaging system manufacturers, more and more of these companies will try and differentiate themselves from the competition by improving the usability of their systems. Those companies which have been in the thermal imaging business longest may prove to have an advantage over their competitors, knowing from their experience the kind of features which maintenance engineers require to carry out a quick and productive thermographic survey.

One unique feature which AGEMA has incorporated on its Thermovision® 550 hand-held infrared cameras is already proving extremely popular amongst customers.

The ability to annotate thermal images with voice messages whilst on a survey is an extremely practical solution which will enable operators to throw away their notebooks and record field information in the most useful place possible – alongside the image to which it relates.

Other features which considerably simplify camera operation and which are now becoming standard functions on many products include automatic set-up of the best image in the camera's field of view, automatic detection of the hottest spot in an image, and differential temperature measurement. Less often found in cameras are functions like dynamic integration – an image enhancement function which selectively averages over stationary objects only and eliminates the smearing effect normally associated with averaging over moving objects.

ACCURACY AND RELIABILITY

A criticism of many of the new FPA systems coming onto the market is their accuracy and reliability. Calibration functions are currently handled in many different ways by system manufacturers. Those systems which automate this function clearly offer major time saving benefits to their customers. Whilst temperature calibration is clearly very important, automation of the lens calibration function will also save the operator important time on a survey should he need to change camera lenses.

Only those cameras which have been developed and manufactured according to international quality standards will provide customers with the assurance they require that the products being presented to them will most likely conform to the various specifications quoted. Above all, these standards provide an indication of the level of quality which customers can expect from such equipment.

RUGGEDNESS

The environment in which some thermal imaging systems must operate demands the most careful system design. Not only should they be extremely rugged and be able to withstand severe jolts, they must be fully sealed against dust and water ingress and preferably to international IP standards. Those FPA systems which include built-in filters which can be switched in or out at the touch of a button offer obvious advantages over those cameras where filters must be added manually.

CHOICE OF SYSTEMS

The use of Focal Plane Array cameras is relatively new within the thermal imaging market. Because of this, many thermal imaging manufacturers are only able to offer customers the choice of one camera for all condition monitoring applications. The performance of such cameras must surely be compromised in those applications for which they have clearly not been designed. Customers must therefore give careful consideration to their individual requirements be they size, weight, image resolution or ease of use etc. and be sure to choose a system which has been designed with these requirements in mind.

REPORT GENERATION

Whilst the improvements in performance provided by FPA cameras are convincing many new customers to adopt infrared inspections techniques, similar improvements in the quality of survey report generators are also having similar effects.

Many system suppliers are now starting to realise the importance of survey reports to maintenance engineers i.e. once they have completed a survey they need to be able to report the results of that survey succinctly and in the shortest possible time. Unfortunately many suppliers of thermal imaging report software suffer from the desire to cram as many features as possible into their software. For the majority of condition monitoring users, many of the analysis functions featured in these products are unnecessary – in many cases a simple analysis of the hottest spot in an image is all that is required!

Those packages which enable the operator to produce a report with minimum effort will clearly prove most popular with customers. The automation of standard procedures will significantly reduce the time required to make a report and in some cases may even allow the camera operator to delegate someone else to do the job whilst he takes the camera away on another survey. This will be an important feature for consultants who need to maximise the utilisation of their cameras to justify their purchase and improve productivity.

CONCLUSION

Since thermal imaging systems first came to be used for commercial applications in the 1960's there have been a number of significant developments which have generated many new important applications. The introduction of Focal Plane Array detectors, however, is a major breakthrough in technology which will change the way people perceive infrared thermography and is already opening up the market to many new customers. Those who weren't convinced the first time about the benefits of infrared condition monitoring, should think again – they may be surprised at what they find!

FOCAL PLANE ARRAYS EXPLAINED

Two of the most important parameters to consider when buying an infrared imaging system are sensitivity and resolution. Sensitivity is a measure of the minimum temperature difference which can be detected or viewed. A system with high sensitivity is able to detect very small temperature differences.

Resolution, sometimes called geometrical or spatial resolution, is a measure of the minimum size which can be resolved by a system. A system with high resolution can detect very small objects.

In order to create an infrared image of an object, a very small infrared sensitive detector cell placed in the focal plane of the camera is required to sense the radiation from one point on the object. Since the detector cell "converts" infrared radiation into an electronic signal, this signal can then be processed by the camera to provide information on the temperature of the object at that point.

The smaller the cell size chosen, the smaller the area of the object which is being sensed. This means that small detector cells must be used to create a thermal imaging system with high spatial resolution. Since the signal level produced by a cell is proportional to its area, this also means that high spatial resolution is achieved at the expense of signal strength (and hence sensitivity).

Another important parameter which affects signal strength is the time the cell senses an object area (known as the integration time). Higher signals are created by those cells which receive radiation over a longer period of time.

In a single element detector system, optical scanning techniques are required to create a thermal picture of the entire surface of an object. Optical mirrors are used to reflect an image of the object on to the detector in such a way that the detector cell is effectively scanning the object horizontally line by line from top to bottom. The integration time in this case is the time required by the detector cell to cover a distance equal to the detector size, typically a few microseconds. The time allowed for the detector cell to scan a line is therefore proportional to the signal strength and subsequent sensitivity of the system. Whilst a longer scanning time would increase sensitivity, this would also have an adverse effect on the image update rate, classifying the system as a slow scan system with low field frequency.

One way to increase the time to scan one line without having a low frequency is to scan many lines at a time. This can be achieved by using many detectors in a column, i.e. a linear array of detectors. By using 160 detectors in an array, the Thermovision® 510 from AGEMA only requires one scan to build up an image of the object. This increases the integration time to the order of milliseconds and has obvious advantages for image quality and sensitivity.

The integration time can be increased yet further by using a matrix of detector cells to receive radiation from an object. Since one cell senses one part of the object all the time, much higher integration times in the order of 15 to 20 milliseconds can be achieved. With such a high integration time, it is now possible to decrease the cell size (and hence increase resolution) without impairing signal quality too much. The Thermovision® 550 from AGEMA uses a matrix array of 320 × 240 detectors.

Thermal imaging systems which use an array of detectors in the focal plane of the thermal imaging system are therefore able to achieve much higher levels of image resolution than single detector scanning systems. Typically, image resolution can be up to four times higher than that achieved by scanning systems.

Figure 1a. Digital voice recording is a unique feature on AGEMA's Thermovision® 550 camera which eliminates the need to take notes on surveys.

Figure1b.

Figure 2a. Camera weight and size are two important features when considering the purchase of a new thermal imaging camera. The Thermovision® 510 from AGEMA weighs less than 1.6kg and can be operated using just one hand.

Figure 2b.

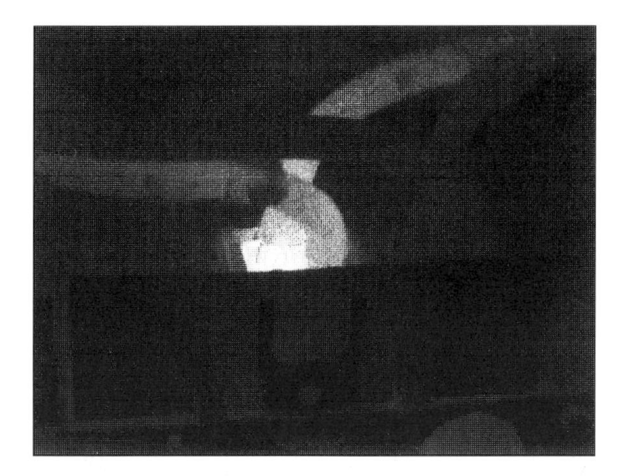

Figure 3. Thermogram of an overheating cable connection taken with the Thermovision® 550 FPA camera from AGEMA.

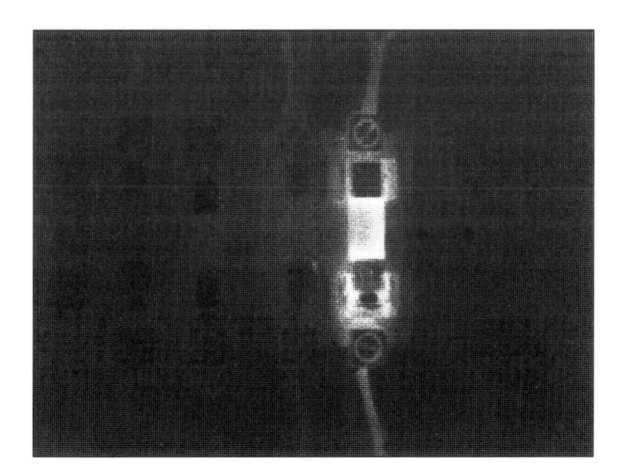

Figure 4. Thermogram of an overheating fuse taken with the Thermovision® 550 FPA camera from AGEMA.

Figure 5. Thermogram of chimney stacks at a petrochemical plant taken with the Thermovision® 550 FPA camera from AGEMA. The surface temperature pattern indicates the condition of refractory bricks.

Figure 6. Thermogram showing the bearing housing on a large fan, taken with the Thermovision® 550 FPA camera from AGEMA.

Figure 7. Thermogram showing the heat loss from a fluid containment vessel, taken with the Thermovision® 550 FPA camera from AGEMA.

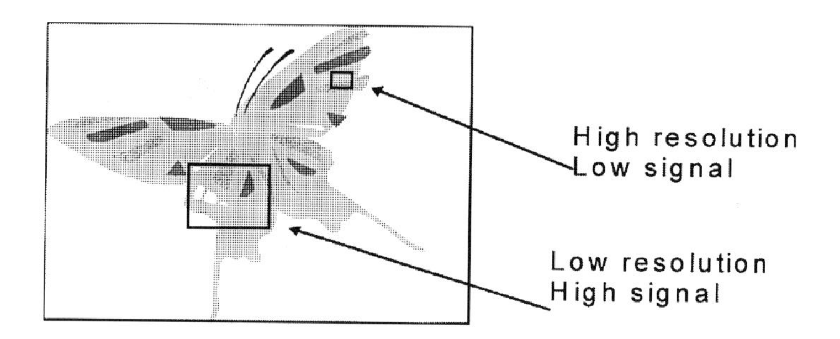

Figure 8a. Graphical representation showing the effects of cell size.

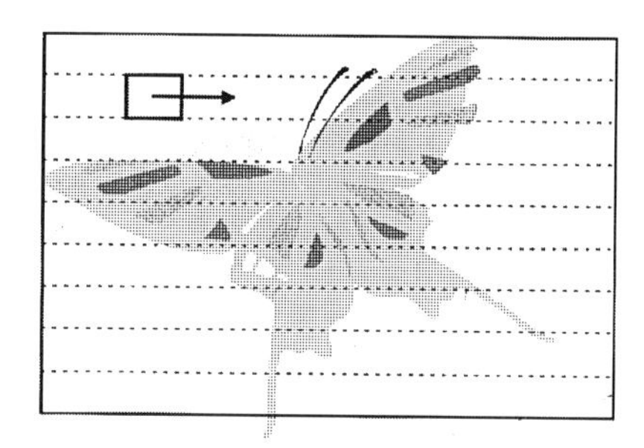

Figure 8b. Graphical representation of a single detector scanning system.

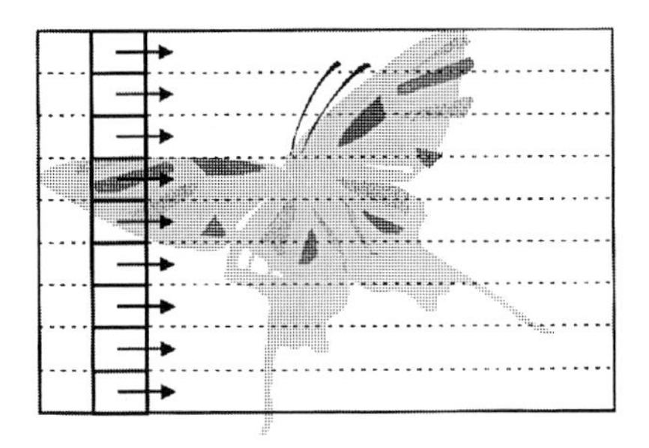

Figure 8c. Graphical representation of a linear array system.

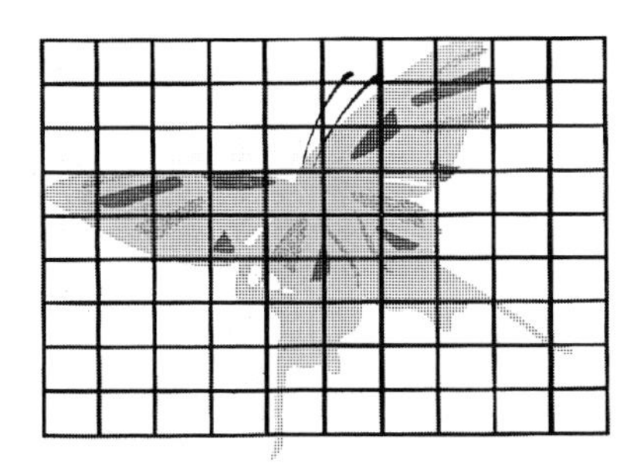

Figure 8d. Graphical representation of a focal plane array system.

Figure 9a. The Thermovision® 550 FPA camera is being used here to make an inspection of the connectors at a substation.

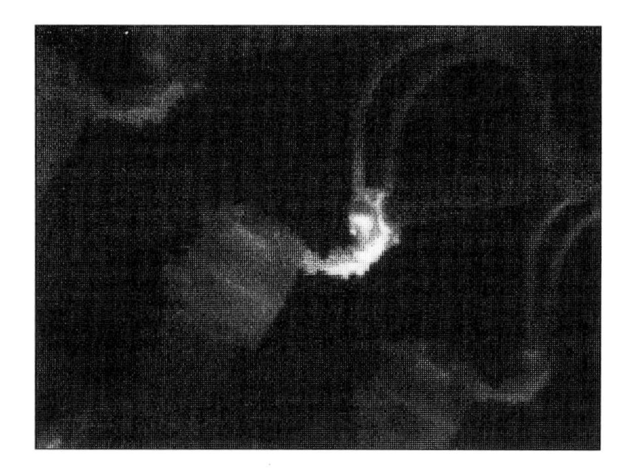

Figure 9b. The thermogram shows that this particular connector is overheating. (Corresponds with Figure 9a).

S472/004/96

Advanced water treatment plants: maintenance – the challenge

M DOWDING CEng, MIMechE
Thames Water Utilities, London, UK

1. HISTORICAL BACKGROUND

In 1990 THAMES WATER was a geographically based organisation with a classical hierarchical structure. The processes used in production were traditional and proven over many decades. Privatisation of the industry coincided with enactment of legislation to improve water quality. As a consequence of the legislation new processes were required particularly for the removal of pesticides and herbicides. Following extensive research it was decided to develop 11 existing surface derived water treatment works and 25 ground water works to Advanced Water Treatment Works status. For the surface derived works this involved at least the introduction of ozone treatment and granular activated carbon filtration. This programme of development would span the decade and was scheduled for completion by the year 2000. Included in the programme was the construction of complete new works at Walton and Grimsbury and the updating of Ashford Common to a reliable output capacity from 540 to 670 Ml/d.

Maintenance policy at the time was based on the concept of Planned Preventive Maintenance. The Computerised Maintenance Management System (CMMS), based on a popular commercially available system, ran on the IBM mainframe computer. Subsequently CMMS was developed "in house" to meet business needs. Standards for maintenance were set by local expert groups. The overall cost of maintenance was monitored and managed.

A programme of cross skilling in the workforce was started coupled with an initiative to harmonise diverse employment conditions within the organisation.

Against this background a review of maintenance policy was completed, linked to other initiatives, to introduce practices to effectively maintain the proposed AWT assets. This presentation outlines the maintenance routines, there implementation and development.

2. CURRENT STATE OF PROGRESS.

Of the 11 surface derived works in the original programme 10 are substantially complete together with 20 of the groundwater works.

The Organisation structure in Thames has been focused onto management of the major business processes. At works level this gives the manager control and accountability for operation and maintenance to predetermined standards. Performance is closely monitored on a "unit cost" basis.

The programme of cross-skilling is drawing to a close. The basic operators has been trained to undertake basic maintenance tasks and the majority of the craft based workforce trained and validated to cross-skilled status (mechanical to electrical and electrical to ICA).

Framework agreements have been put in place for the procurement of assets such as pumps, valve and switchgear. This enables maintenance routines to be refined for a rationalised asset base and spares stockholding to be optimised. Procurement of spares electronically via an interface between the CMMS and the purchasing system has been developed and is currently being introduced.

Technical standards for maintenance in the form of Best Practice for specific assets has been established and applied to maintenance routines generally on the basis of risk of failure to the Process. The Best Practice initiative is intended to focus maintenance onto consistent technical standards across the whole of Thames Water. Best Practice for individual AWT processes is evolving based upon manufacturers original recommendations and experience gained in operation of new plants.

Maintenance policy is being focused on the principles of Reliability Centred Maintenance. The following evolving linked modules form a Technical Standard for maintenance are key to this policy:-

1. Design for Maintenance.
2. Process Failure Analysis.
3. Process Failure Review.
4. CMMS.

Design for Maintenance focuses the Thames Engineering Constancy on the need to consider, in the design process, procedures to ensure that current maintenance Best Practice can be applied to evolving plant design. New Best Practice will be evolved where necessary.

Process Failure Analysis links risk, process reliability and the appropriate maintenance schedule. The maintenance schedule is a balanced mix of planned, corrective, detective and condition based maintenance. Detective and condition based maintenance trigger designed corrective based maintenance. Where an appropriate maintenance schedule cannot be determined to prevent process failure design modifications are considered.

Process Failure Review is an ongoing activity whereby, on a regular basis, process failures are reviewed to determine the effectiveness of maintenance and review the maintenance schedule and Best Practice where necessary. The objective is to improve reliability and cost effectiveness. Design modifications can be evolved where a maintenance routine cannot prevent process failure.

The fourth module (CMMS) covers the Code of Practice for application of the Computerised Maintenance Management System.

Maintenance activity in managed units at individual works or in areas is applied to a generic model. Condition monitoring, basic tasks, inspections, routine checks are dedicated to rounds. Planned corrective and preventive maintenance are resourced to cross-skilled teams dedicated to each unit or to appropriate contractors.

All activities are implemented through quality procedures compliant with the Thames Quality Standard.

3. FUTURE DEVELOPMENTS

The in house developed CMMS is no longer capable of realistically managing the demands of the maintenance operation development to upgrade the system to operate within the future Thames IT strategy is not beneficial to the business. A project is being evolved to replace the system with a "state of the art" commercially available package.

Key user requirements for the new system will equal the current functionality together with enhancement to introduce the hand-held computer terminals for rounds management and facilities for maintenance planning. An interface with the SCADA management system is envisaged to enable condition monitoring techniques to be developed.

The current budget for all company wide plant maintenance activity is £35m per annum. Significant cost savings are achievable in this area and application of this philosophy to all maintenance activities is planned to coincide with introduction of the new CMMS.

Developments in the overall IT system will facilitate the introduction of electronic data and record management to supplement the CMMS.

S472/005/96

Early rewards following the application of condition monitoring

R A WYNNE MEng & Man, CEng, MIMechE
Albright and Wilson Limited, Wolverhampton, UK

Introduction

Albright & Wilson (UK) Ltd (A&W) is a multi national chemical company with its technology centred around phosphorus chemistry.

The company was founded in 1851 by Mr Arthur Albright and Mr Johnathon Wilson (both Quakers) manufacturing matches using elemental Red Phosphorus.

Nowadays it enjoys an international reputation and its portfolio includes many components for household utility products such as toothpaste, washing powder, shampoo etc.

A basic organogram indicating the authors position within it may be found in Fig. 1.

Basis For Change

Albright & Wilson (UK) Ltd (Oldbury) was typical of most companies, being predominantly reactive in its approach to maintenance. The engineering divisions pride was its ability to undertake rapid, quality emergency maintenance.

In the 1970's this approach was recognised as not being cost effective and was supplemented with a planned shutdown programme. Every year plants would be shut down for their annual overhaul. Unfortunately during these outage periods numerous items were stripped down and parts renewed unnecessarily i.e. repairing the unbroken.

Following studies in civil aircraft it has been shown that although the figures below are not directly transferable to the chemical industry, they are however comparable.

The traditional 'bath tub' curve indicating component life cycle is only truly applicable to approximately 4% of the population. Modern practice involves the 'J Curve' (see Fig. 2) which has been found to be applicable to $\frac{2}{3}$ of the population. Hence, in the past, units were being stripped down and overhauled soon after they had achieved their optimum operating state.

This knowledge coupled with the awareness of high maintenance costs and low plant occupancy initiated the setting up of a condition based monitoring program.

The report concentrates primarily on the vibrational analysis aspects of CM. However, at Oldbury other aspects include :

i) Ultrasonic inspections of vessels and pipework
ii) Spark testing of glass and plastic lined vessels.
iii) Lubrication analysis.

Condition Monitoring (CM)

The Initial Stages

Prior to the introduction of CM it was imperative that management were clear of the direction and implications of the scheme and were totally supportive of it. This happened to be the case at Oldbury and so the monies and resource needed to launch the scheme were readily obtained.

Also it was necessary to cite relevant examples to support the introduction of CM. Typical of such examples was that of a failed extraction fan. This incident occurred two weeks following the maintenance shutdown of a PHOSPHINE plant.

It resulted in the total shutdown of the plant with attendant costs :-

i)	Bearing renewals (cause of failure)	£400
ii)	Lost production	£160,000
iii)	Major disruption to the business	

The above incident provided a clear message that the cost of mechanical spares can be infinitesimally small when compared to the cost of lost production and possible larger consequences of lost opportunities. Hence, a structured introduction of CM began in early 1993.

Preparatory Work

To assist in the introduction of CM cost effectively it was decided to carry out a criticality study. This basically involved analysis of plant and machinery to determine all cost implications of various item failures. The team seconded to conduct the study comprised :

Engineer
CM Technician with assistance from Engineering Supervisors

Following this study it was then possible to prioritise equipment to help achieve early success and gain momentum.

Also, during this period a Condition Monitoring Technician (CMT) was appointed and trained accordingly.

The Technician then worked through the program resulting from the criticality study obtaining all relevant information on each item in turn. He then generated the data base from which he created ordered testing routes within each plant.

Obviously, the collection of data for each plant in turn started immediately following the creation of its route. Thus the database and unit histories expanded together.

Data Collection

The CMT selected a particular route from the P.C. and down loaded all relevant information into the hand held data collector. He then completed his analysis obtaining horizontal, axial and vertical vibration measurements at each bearing location. On his return, all information was loaded on to the P.C. and an update of the routes history obtained. Any readings which were greater than the alarm point were highlighted by the software automatically.

Analysis

All machines within Oldbury Works operate within the 600 - 60,000 rpm range and the two prime areas of interest are velocity and spike energy.

Velocity provides an indication of vibration severity and spike energy assists in determining the location of the vibration. (See FIG. 3).

As each type of vibration has its own individual characteristics they will appear at different frequency bands. Hence the mechanical problem can be identified by analysing the velocity spectrum.

Within 8 months (end 1993) the program was completed and at this stage to date each of the units has been subjected to at least one analysis. The inspection interval during the first few months was weekly in order that results would be obtained at earliest opportunity. Subsequently the interval was extended to monthly.

The IRD system is used as an on-request analyser to identify the cause of vibration whether from bearing damage, misalignment, imbalance etc. However, its core use is that of trending measurements over time. From this the inspector can gain an understanding of the machine and identify any pertinent changes far sooner than an individuals senses can.

The system has operated for the past three years with new equipment being introduced at the project stage.

Early Rewards

With any cost improvement exercise an accountable method of recording the cost benefits must be employed. The forum that has been used at Oldbury consistently over the past three years is known as the "Cost of Quality" (CoQ) program.

Essentially downtime is recorded monthly onto a Lotus 1-2-3 spreadsheet by the Production Supervisor. The figures are divided into scheduled and unscheduled with firm guidelines indicating which maintenance activity falls into which category.

Consequently misinterpretation is eliminated and the figures reflect the true picture.

The "CoQ" results over the past three years are outlined below:

	Cost of Quality Project		
	Phase 1	Phase 2	Phase 3
Yearly Saving (Target)	£120K	£108K	£97K
Yearly Saving (Actual)	£568K	£296K	£1013K (YTD)
% of target	473%	273%	1044%
Cost to employ	£20K	£20K	£20K
Net yearly saving	£548K	£275K	£993K

As can be seen from the results, in the first year (phase 1) a great saving was achieved primarily due to the change in maintenance philosophy i.e. the introduction of CM.

The second year (phase 2) was still a good year even though the baseline from which the savings are measured had improved due to last years success.

This year (phase 3) is by far the most encouraging. The information on each item has improved, our historical trends are far more comprehensive and the inspector has a far greater understanding of the condition of each item.

Whilst the above savings may be, as stated, due to CM the contribution of other site initiatives cannot be ignored. These initiatives, which have led to a leaner organisation and an improved maintenance function also created the atmosphere/culture which aided the CBM program.

Latest Developments and The Way Forward

The company is currently pursuing a Reliability Centred Maintenance (RCM) approach and as a directive from management all new capital projects are subject to an RCM study at the design stage.

Such an RCM study is effectively a Hazard Operability (HAZOP) study for maintenance. It does not define a correct maintenance strategy i.e. planned preventative, but it does generate "horses for courses" and therefore examines each units failure pattern and offers the correct maintenance philosophy for that unit.

Learning Points

i) It is imperative that management issue clear and unambiguous direction for the initiative by the issuing of a 'Mission' statement.

ii) This 'Mission' statement must be cascaded down to the workforce who are then aware of the need for their support when requested.

iii) Initial costs of setting up condition monitoring may in isolation appear high. However, after only a short time CM generates cost savings far in excess of initial costs.

iv) As the introduction of CM is to improve efficiency in maintenance it is imperative that an accurate accounting system is in place.

v) Condition monitoring as applied to rotating equipment must be accompanied by an effective lubrication program.

About the Author

The author is 27 years of age and is employed by Albright & Wilson (UK) Limited, at their Oldbury works. Having graduated from Birmingham University in 1991 with a first class honours degree in Mechanical Engineering, Manufacturing and Management, he joined A&W Ltd as a Project Engineer. During his five years with A&W he has held numerous positions dealing with project, development, maintenance engineering and has also managed a newly appointed maintenance planning department. His current position is that of Area Plant Engineer, responsible for 8 production units vital to the successful operation of the Works.

FIG. 1.

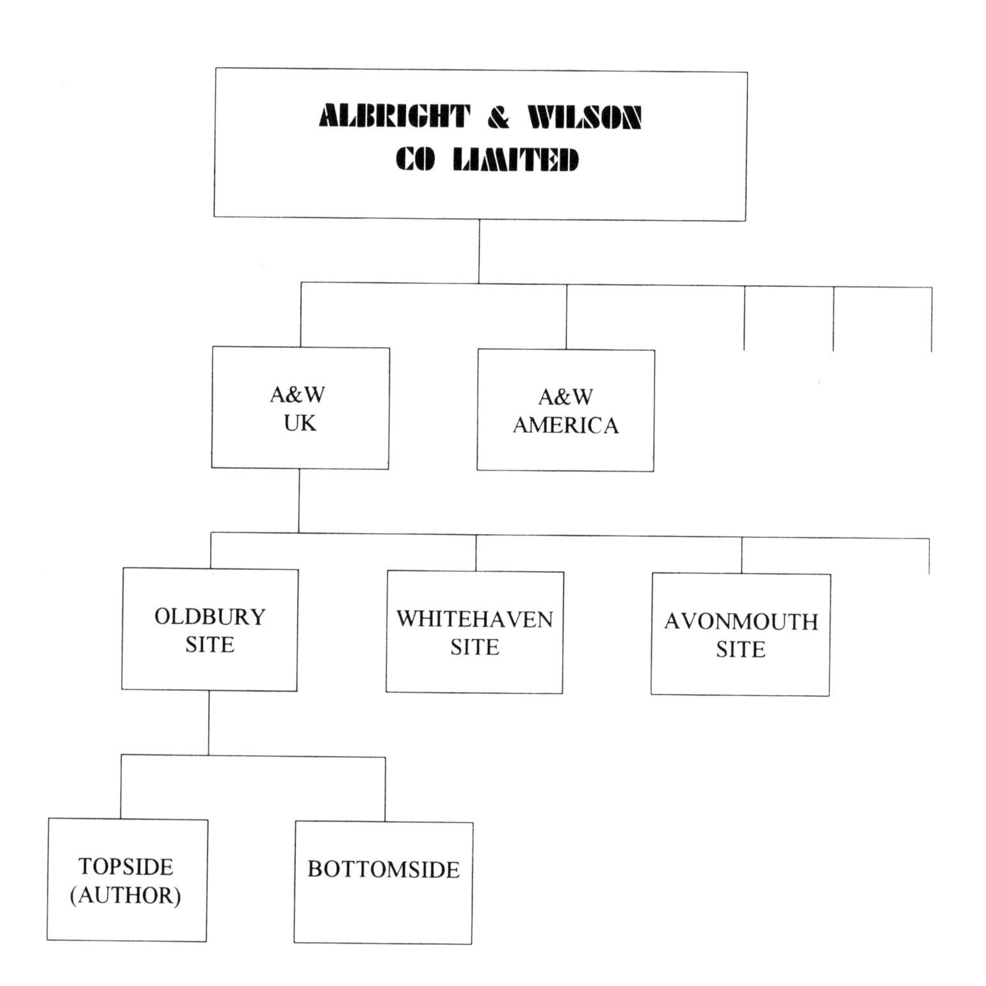

ALBRIGHT & WILSON
CO LIMITED
A&W
UK
A&W
AMERICA
OLDBURY
SITE
WHITEHAVEN
SITE
AVONMOUTH
SITE
TOPSIDE
(AUTHOR)
BOTTOMSIDE

Fig.2

BATH TUB VS THE J CURVE

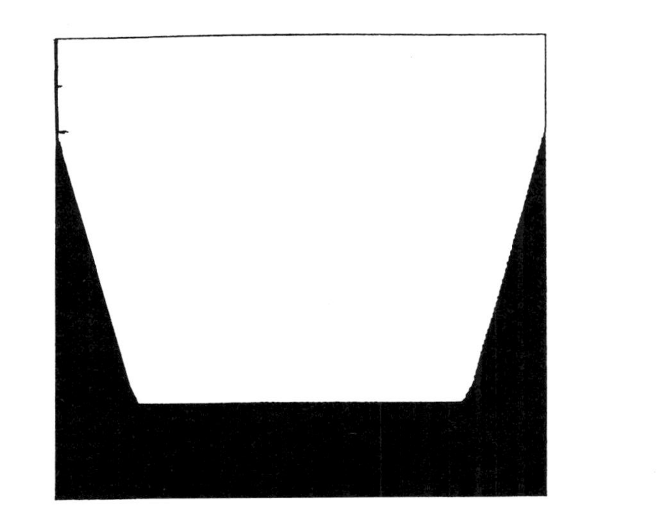

FIG. 3

1.Velocity Trend Graph

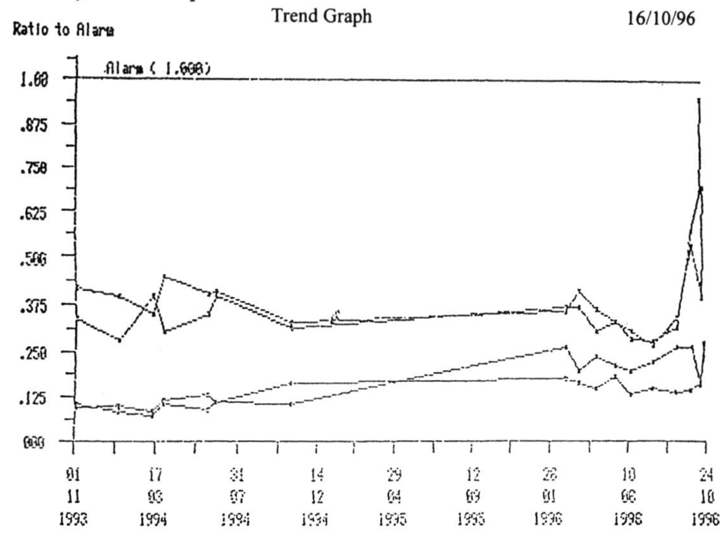

2. Velocity Spectrum

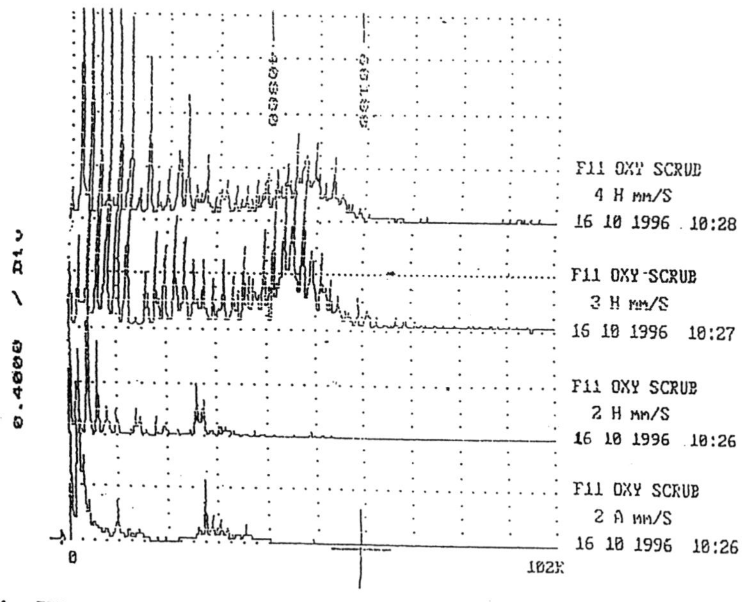

Fig. 3 (Cont).

Vibration Identification Chart (amplitude, frequency, phase)

Cause	Amplitude	Frequency	Phase	Remarks
Unbalance	Proportional to unbalance. Largest in radial Direction	1 x RPM	Single reference Mark- stable repeatable	Most common cause of vibration
Misalignment Couplings or Bearings and Bent Shaft	Large in axial direction. 50% or more of radial vibration	1 x RPM usual 2 & 3 x RPM sometimes	Single Double or Triple	Best found by appearance of large axial vibration. Use dial indicators or other method for positive diagnosis. If sleeve bearing machine and no coupling misalignment, balance the rotor.
Bad Bearings Anti-Friction Type	Unsteady-use velocity, acceleration, and Spike Energy Measurements	Very high Several times RPM	Erratic- Multiple Marks	Bearing responsible most likely the one nearest point of largest high frequency vibration. Spike Energy Measurements recommended when analyzing bearing failures.
Eccentric Journals	Usually not large	1 x RPM	Single Mark	If on gears largest vibration in line with gear centers. If on motor or generator vibration disappears when power is turned off. If on pump or blower attempt to balance.
Bad Gears or Gear noise	Low-Use Velocity, Acceleration and spike Energy Measurements	Very High Gear Teeth times RPM	Erratic- Multiple Marks	Velocity, Acceleration, and Spike Energy measurements recommended when analyzing gear problems. Analyze higher orders and sideband frequencies.
Mechanical Looseness	Sometimes Erratic	2 x RPM	Two Reference Marks Slightly Erratic	Usually accompanied by unbalance and/or Misalignment
Bad Drive Belts	Erratic or Pulsing	1,2,3, & 4 x RPM of Belts	One or Two depending on Frequency, Usually Unsteady	Strobe Light best tool to freeze faulty Belt
Electrical	Disappears when power is turned off	1 x RPM or 1 or 2 x synchronous frequency	Single or Rotating Double Mark	If vibration amplitude drops off instantly when power is turned off cause is electrical. Mechanical and electrical problems will produce "beats".
Aerodynamic or Hydraulic Forces	Can be large in the axial direction	1 x RPM or Number of blades on fan or impeller x RPM	Multiple Marks	Rare as cause of trouble except in cases of resonance.
Reciprocating Forces	Higher in line with motion	1,2 & Higher orders x RPM	Multiple Marks	Inherent in reciprocating machines, can only be reduced by design changes or isolation.

S472/006/96

Recent developments in off-line and on-line vibration monitoring equipment

B FINCH BSc, CEng, MIMechE
Bently Nevada (UK) Limited, Warrington, UK

MEASUREMENT PHILOSOPHY

When setting out to monitor vibration on a rotating machine, it must first be determined whether to monitor shaft motion directly, or to measure the resulting bearing housing vibration. On machines with rolling element bearings, the bearing stiffness is high and internal clearances are small, and so there is little relative motion between the shaft and the bearing housing. Hence, with rolling element bearings it is usual to measure bearing housing vibration.

On machines with sleeve bearings however, the decision whether to monitor bearing housing motion or shaft motion relative to the bearing is not so straightforward, and it is necessary to take into account bearing oil film stiffness, casing/rotor mass ratio, bearing housing stiffness etc. On machines with relatively light rotors, moderate oil film stiffness and massive rigid casings, such as is normally the case with centrifugal compressors, then shaft relative motion is the preferred measurement. Bearing housing measurements would be preferred on sleeve bearing machines with heavy rotors, high oil film stiffness and flexible supports. On some machines, such as large turbine generators mounted on concrete foundations, to monitor the vibration fully, both casing and shaft relative measurements should be taken.

It should be noted that there is no fixed relationship between the shaft relative motion and the corresponding bearing housing motion on a sleeve bearing machine. A change in shaft relative motion may not be accompanied by a significant change in bearing housing vibration.

MEASUREMENT UNITS

The three vibration measurement parameters are displacement, velocity and acceleration. The relationship between them is:

Velocity = 2 x ∏ x frequency(Hz) x displacement

Acceleration = 2 x ∏ x frequency(Hz) x velocity

Hence, for a fixed displacement, the corresponding velocity level increases linearly with increasing frequency, whilst the corresponding acceleration value increases with the square of frequency.

Shaft relative motion is normally measured in terms of peak to peak displacement because those units give the total shaft motion within the bearing clearance.

Where bearing housing vibration is concerned, machines of the same type and condition tend to vibrate at constant velocity, independent of speed, and velocity levels have been found to be good general indicators of machine condition. Velocity is generally measured in terms of rms or peak values. (It should be remembered that the relationship: peak = 1.414 x rms is true for a sine wave, but the ratio can be considerably higher for a complex signal).

As acceleration is highest at high frequencies, in applications where high frequency vibration is of interest, such as at tooth mesh frequencies on gearboxes (which can be at several kHz), acceleration is the preferred parameter. Either peak or rms values are used. A true peak measurement is more sensitive to changes in machine condition than is an rms measurement.

SENSOR TYPES

Non-contacting eddy current probes are normally used to measure shaft displacement and position. For casing vibration measurements either a velocity transducer or accelerometer is used. The accelerometer output can be integrated to give vibration in terms of velocity.

MONITORING PARAMETERS

On smaller general purpose machines, which often have rolling element bearings, the vibration parameters which are trended are normally direct (i.e. unfiltered) levels, plus frequency spectrum information. This is usually adequate to indicate the onset of problems.

On larger machines which are critical to production, and generally have sleeve bearings, additional vibration parameters which are monitored are 1 times (1X) running speed and 2 times (2X) running speed vibration vectors (i.e. magnitude and phase angle of the vibration components), along with the shape or form of the motion. In some circumstances, higher order vectors are also monitored. Transient (runup and rundown) information is often recorded. Additional information is available from eddy current probes in the form of shaft centreline position. This enables shaft position changes to be monitored in sleeve bearings.

Figure lA shows steady state data from a machine in good running condition. It includes unfiltered and filtered orbits, a spectrum plot, lx vector trended against time, 1X vector trended in Polar format and a Waterfall Spectrum plot. Figure 1 B shows transient plots in Bode, Polar, Cascade Spectrum and shaft centreline formats. The first balance resonance (or critical speed) can be seen at 3400 rpm.

OFF LINE MONITORING SYSTEM - PORTABLE DATA COLLECTOR

It is usual for spared non-critical machines to have no permanently installed vibration monitoring systems. These machines are often monitored using a portable data collector, which an operator carries round a predetermined route, taking vibration data from a series of machines. The data stored in the collector is then down-loaded into a computer, which flags any machine where the vibration has gone outside preset values. Additionally, if suitable software is installed, the computer will identify the source of the problem.

This system works well, and because the operator comes into close contact with the machines, any developing problems, such as leaks, can be identified. However, the portable data collector system is labour intensive, and the frequency of data collection must be greater than the time-to-failure of a machine, if problems are to be detected before damage occurs.

The Waterfall Spectrum Plot in Figure 2 shows a series of spectra from a deep groove rolling element bearing. Readings were taken on a weekly basis and it can be seen that the period between the first signs of the generation of significant vibration components (in this case at rolling element outer race passing frequency) to the time that the machine was shutdown to replace the badly damaged bearing, was less than two weeks. If sampling had been carried out less frequently, the failure could have been missed.

OFF-LINE MONITORING SYSTEM - MULTI-CHANNEL TRANSIENT DATA COLLECTOR

This system is normally used for trouble shooting on important machines which generally have permanent vibration monitoring installed. The monitors provide 'Alert' and 'Danger' alarms to operators as machine vibration levels change. Often analogue outputs from the monitor, proportional to direct vibration levels, are trended by the DCS system.

In order to diagnose problems on these machines it is often useful to monitor run-up and run-down data. To do this, if an on-line system is not installed, a portable multi-channel data collector, which is triggered from a keyphasor™ (once per revolution) reference pulse, samples the raw vibration signals from the monitors. (If transducers are not fitted, magnetically attached casing transducers and a temporary optical keyphasor can be used). The unit then tracks the machine running speed and stores vibration vectors, waveforms and frequency information (along with gap voltages, if eddy current probes are fitted) during the machine start-up, steady state and shut-down.

The transient data enables the frequency and synchronous amplification factor of balance resonances (aka critical speeds) to be determined.

As an example, Figure 3 shows run down data from a large motor which had high levels of 1X running speed vibration. Initially the cause of the high vibration was not known, but from the run down plot it can be seen that the machine was running at the first balance resonance speed (aka critical speed), thus rendering it sensitive to relatively small changes in balance. The rundown information was captured automatically when the machine tripped.

ON-LINE STEADY STATE GENERAL PURPOSE MACHINERY PERIODIC MONITORING

This system is an alternative to using the walk round data collector. It is more capital intensive to start up, but has lower running costs. One transducer and interface module with a unique address is permanently attached to each machine bearing - see Figure 4. The transducers are linked together and back to the data acquisition computer by two twisted cable pair daisy chains, which minimises installation costs. The computer accesses the signal from each transducer in turn and trends the information. The computer can be programmed to provide 'alarm' indications on various vibration parameters.

An advantage of this system is that each measurement position is checked more often (at least three times a day) and more reliably than is the case with a portable data collector. The computer can be programmed to look at problem machines more frequently. With the use of networks and modems the incoming and trended data can be examined by engineers either in the plant or at remote locations.

ON-LINE CONTINUOUS MONITORING OF CRITICAL MACHINERY UNDER STEADY STATE AND TRANSIENT CONDITIONS

On critical plant, it is normal practice to install permanent monitoring systems. With continuous on-line monitoring, the vibration signals from the permanently installed transducers are collected, processed and buffered locally before being fed back to the data acquisition computer.

Runup/rundown and steady state data are stored and trended by the computer, effectively providing a continuous record of the vibration behaviour of the machine - see Figure 5. With this system, transient Bode, Polar, Cascade Spectrum, Shaft Centreline and other plots can be produced for each transducer on each machine, even if all the machines monitored are running up or down simultaneously. All the steady state vibration parameters are trended to provide an on-going record of the machine behaviour, and deviations from the norm can be set to trigger alarms. The data acquisition computer can be networked to other computers so that incoming vibration data can be observed by engineers at various locations in the plant or remotely via a modem. When a problem begins to develop, the immediate availability of the historic data enables rapid informed decisions to be made.

EXAMPLES OF MACHINERY PROBLEMS

Rotor radial rub (1)

Figure 6 shows a series of unfiltered shaft orbits taken as a centrifugal compressor was started up after an overhaul, during which new seals had been fitted. The first balance resonance was above 4500 rpm. The once per revolution keyphasor blank-bright marker is shown on the orbits. The vibration amplitude started to increase at 1800 rpm, due to a radial rub which heated the rotor locally, causing a shaft bow to occur. The bow, in turn, caused the rub to become more severe. At first the orbit remained reasonably elliptical, but soon after the machine was tripped, the motion became more triangular, indicating that impacts, which were sufficient to deflect the shaft motion, were occurring three times per revolution. As the machine ran down, the motion became more triangular still, as the impacts became more severe, until the rubs were heavy enough to cause reverse precession at the points of impact (1921 rpm), producing three lobes on the orbit. The final orbit at 849 rpm was considerably larger than the 1050 rpm orbit on the run-up, showing the presence of the shaft bow due to the heating effects of the rub.

Observation of the orbits during the run enabled an instant diagnosis to be made. The solution here was to wait for some 60 minutes until the rotor had cooled and straightened and then run the machine up again. The second start was trouble free as the rub had increased the clearance in the abradable seals enough to prevent the rub recurring.

Rotor radial rub (2)

On a large steam turbine it was seen over a period of several years that the IP (Intermediate Pressure) shaft 1X vibration vector cycled continuously following reductions in Generator load - see Figure 7. It was concluded that the accompanying reduction in IP inlet steam temperature caused distortion to occur at the IP cylinder, which produced a radial rotor/stator rub. This machine was operating below a resonance, but sufficiently near to it for the vibration phase angle (ie the rotor deflection) to be lagging the location of the unbalance on the rotor.

Here again the rub heated the rotor, causing it to bow radially, effectively producing an additional unbalance, which lagged the residual unbalance. The location of the resultant combined unbalance also lagged the position of the original residual unbalance slightly, causing the vibration phase angle to increase. This increase caused the position of the rub on the rotor to become more lagging still, and thus continual slow cycling of the vibration phase angle occurred until the rub cleared as the load was increased again. The severity of the vibration vector swings had decreased progressively over the years, indicating that the rubbing was causing wear which gradually increased the internal clearances.

Rotor radial rub (3)

On a machine where a balance resonance (aka critical) speed lies just below half running speed and a radial rub occurs, the effect of a rub can be two-fold. It effectively increases the net system stiffness, increasing the frequency of the resonance to half running speed, and the impetus imparted to the shaft due to the rub re-excites the modified resonance at exactly half running speed. The spectrum plot in Figure 8, from a large steam turbine HP rotor, shows the presence of a vibration component at exactly half running speed due to a radial rub, whilst the orbit plot shows that the impact occurred at the 3 o'clock position on the orbit. By definition, the keyphasor mark aprs once during each shaft revolution, so it can be seen that, following the impact, the shaft rebounded away and made two revolutions before the impact recurred. Re-aligning the rotor 0.25 mm (10 inch x 10^{-3}) to the left, with respect to the casing, eliminated the problem.

MISALIGNMENT

Figure 9 shows a shaft orbit at a bearing which had an excessive downward load, due to the bearing being misaligned high. The shaft was forced hard down in the bottom of the bearing, constraining the motion to follow the contour of the bearing. The shaft motion was initially high (the major axis motion was 230 microns pk-pk), but the levels decreased and the orbit shape became more elliptical as bearing wear (which was trended continuously from the eddy current probe gap voltages) took place. Ultimately some 2 mm of bearing wear occurred (out of a total white metal thickness of 2.5 mm) during the running period.

FLUID INDUCED INSTABILITY

The vibration level on a centrifugal compressor running at 10800 rpm increased for no apparent reason, forcing a reduction in machine speed (and consequent loss of production), to keep the vibration down to acceptable values. Analysis showed that the increase was caused by a frequency component appearing at 4600 cpm, well below half running speed - see Figure 10. The component was lower in frequency than would be expected for oil whirl (c. 47% of running speed) and a check back to previous runup data for the machine showed that the 4600 cpm component corresponded to the frequency of the first balance resonance, indicating that a fluid induced instability was re-exciting the first balance resonance - a phenomenon usually referred to as oil whip. Oil whirl and whip often occur in lightly loaded cylindrical bearings. This machine had tilt pad bearings fitted, which, if correctly sized, cannot go unstable. In this case, it was suspected that one of the compressor floating ring seals had locked up and was behaving as a lightly loaded cylindrical bearing. Various checks were carried out and it was found that the subsynchronous vibration disappeared if the seal oil inlet temperature was reduced. This enabled the machine to continue in production, with the seal being replaced later, at a convenient shutdown.

FUTURE DEVELOPMENTS

Process and power industries are under increasing pressure to reduce costs and increase operating efficiencies. A primary vehicle for reducing costs is the improved management of machines. Hence, more complete information is needed, in an easily accessible form so that effective and timely decisions can be made.

Using digital communication links, future computerised monitoring systems must be able to pass vibration and process information to the plant multipurpose information network, and also accept process data from the DCS or process computer, thereby enabling the direct correlation of process data with the vibration data.

Plants which have machinery engineers with in-depth diagnostic experience are becoming fewer. Automated diagnostics can be provided by knowledge-based programmes which may be resident on the acquisition computer, one of the networked computers, or on a computer at a remote location. The programme acquires data directly from the on-line system and produces audits which identify problems as they arise. The knowledge-based system checks the various trended parameters before giving a conclusion. Shaft bow, High synchronous vibration, Fluid induced instability, Radial pre-load forces, Vector change, Rotor rub, Loose rotating parts, Compressor surge and Motor non-uniform air gap can be identified.

Alternatively, diagnostics can be carried out remotely by experienced engineers. By use of modems and display software (and knowledge of the machine), the on-line system can be accessed from anywhere in the world and a report produced on the machine condition. Whilst this is not new, the use of remote diagnostics is expected to increase sharply, as it can eliminate the costs associated with an engineer travelling to site. A further advantage is that with access to experienced diagnostic engineers world-wide, expertise is available twenty-four hours a day.

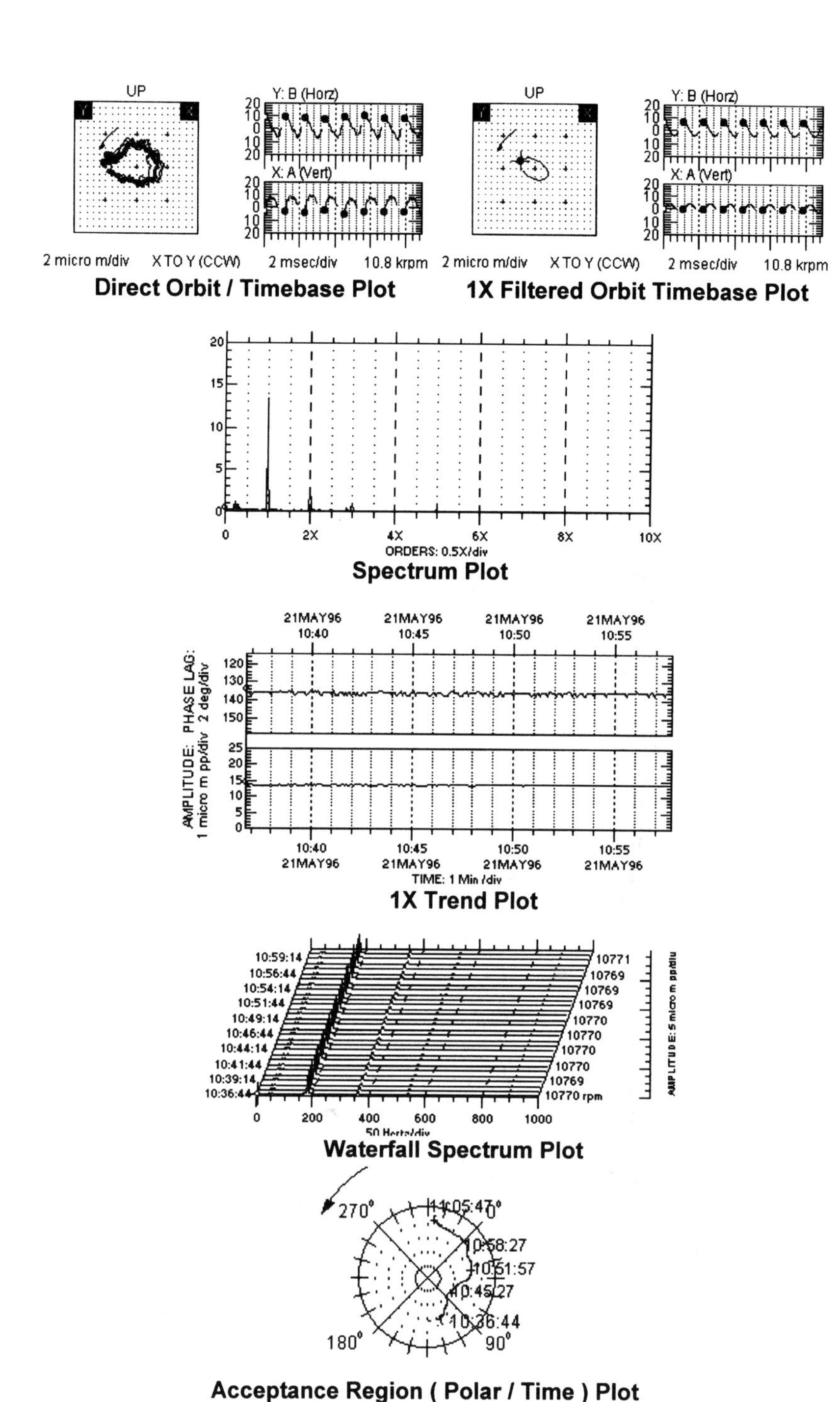

Figure 1A - Typical steady state data and trend plots.

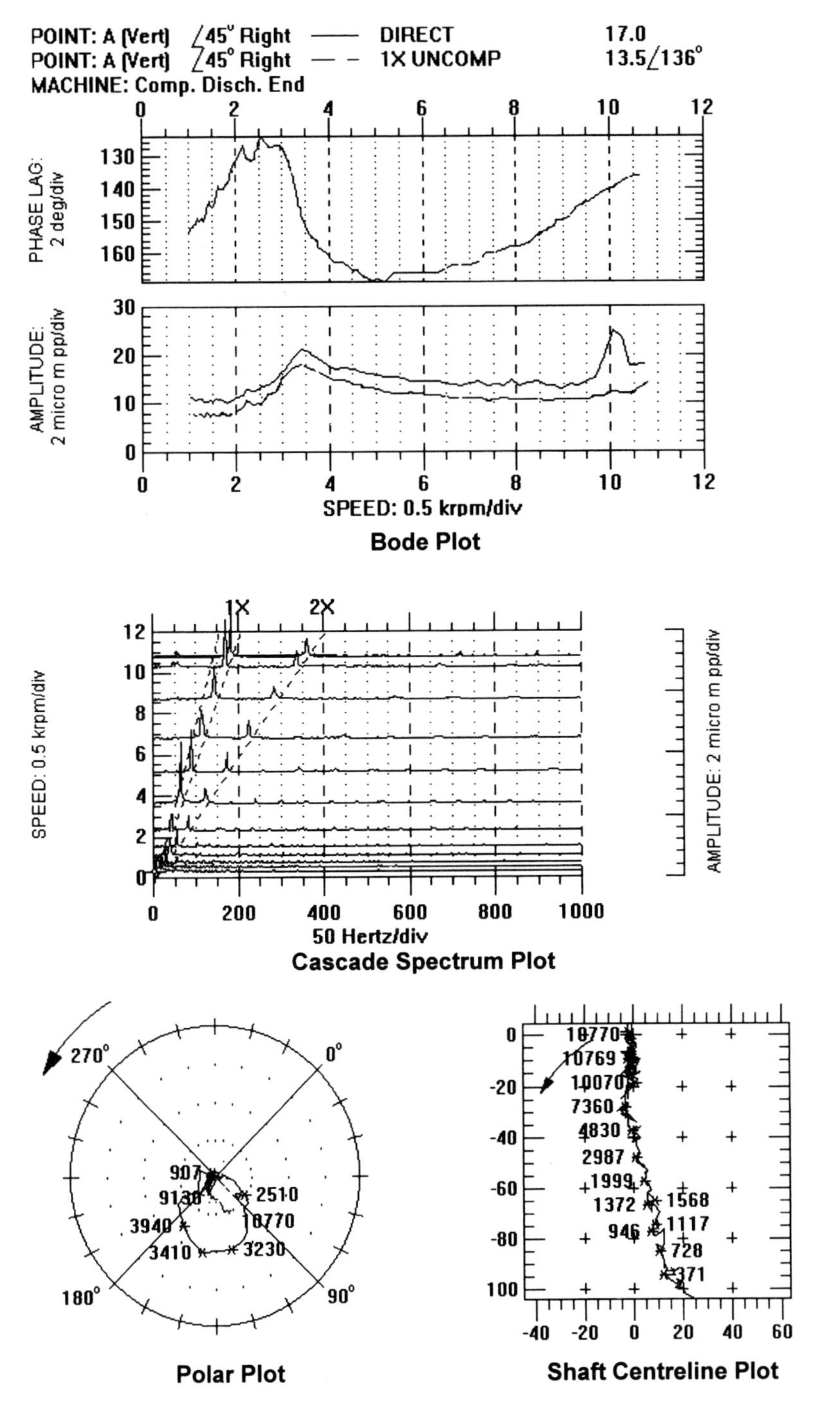

Figure 1B - Typical transient data plots.

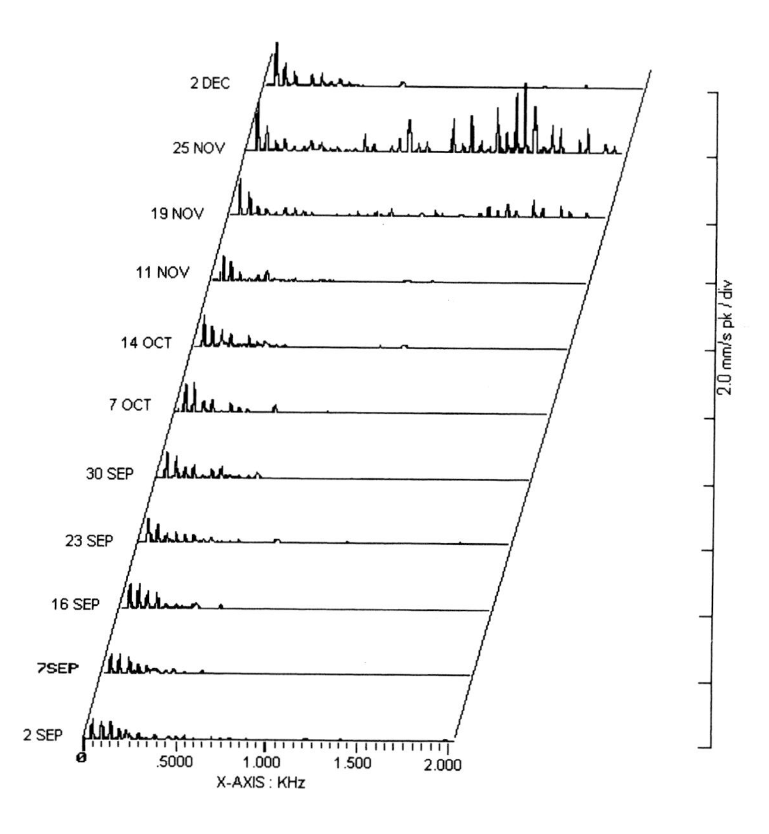

Figure 2 - Waterfall spectrum plot using portable data collector

Deep groove rolling element bearing. Velocity readings taken weekly. Note the onset of outer race failure frequencies on 19 November. The machine was shut down and the badly damaged bearing was replaced following the 25 November readings.

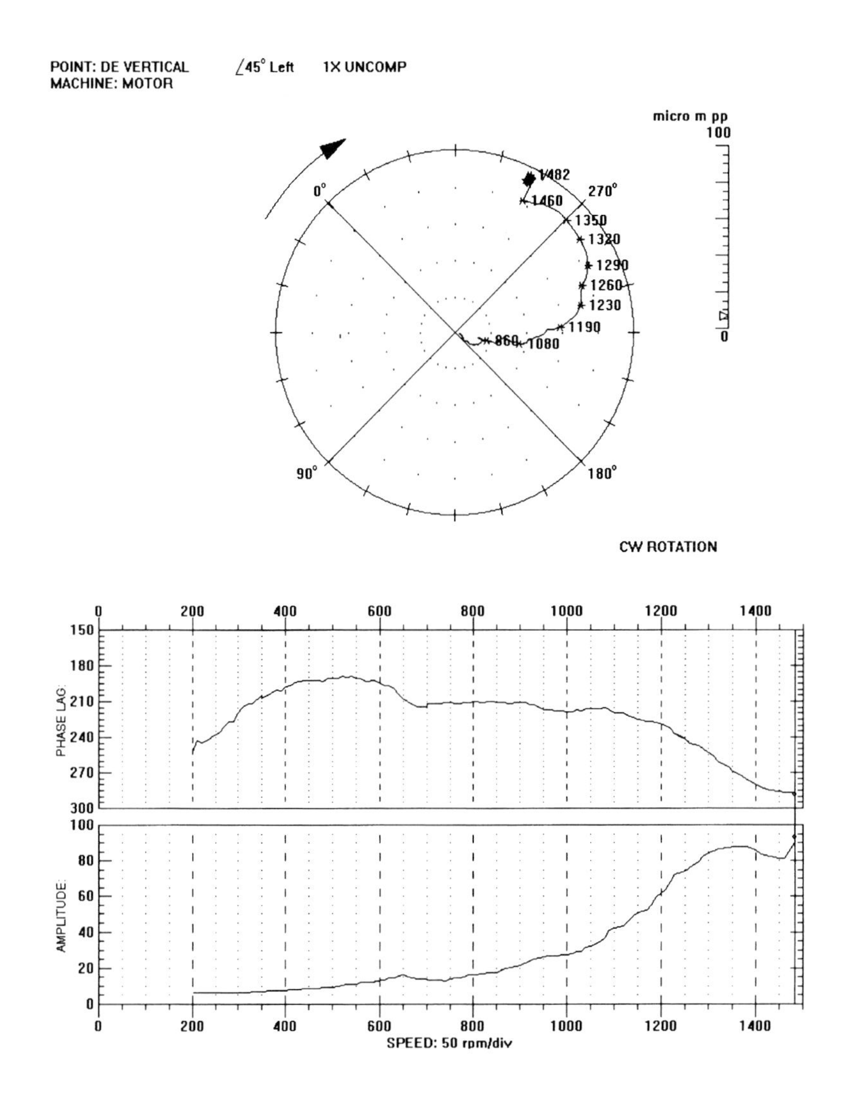

Figure 3 - Run down plots from a large motor, showing the first balance resonance near to running speed.

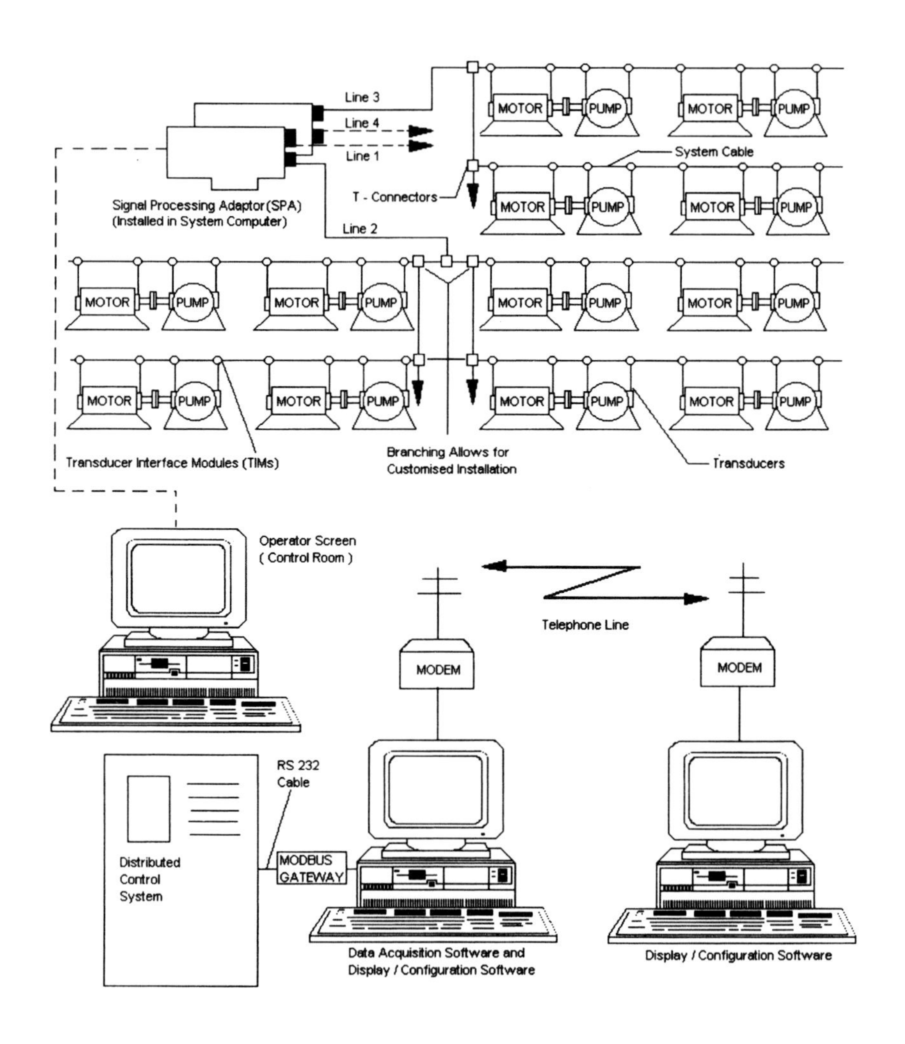

Figure 4 - On line steady state general purpose machinery monitoring system.

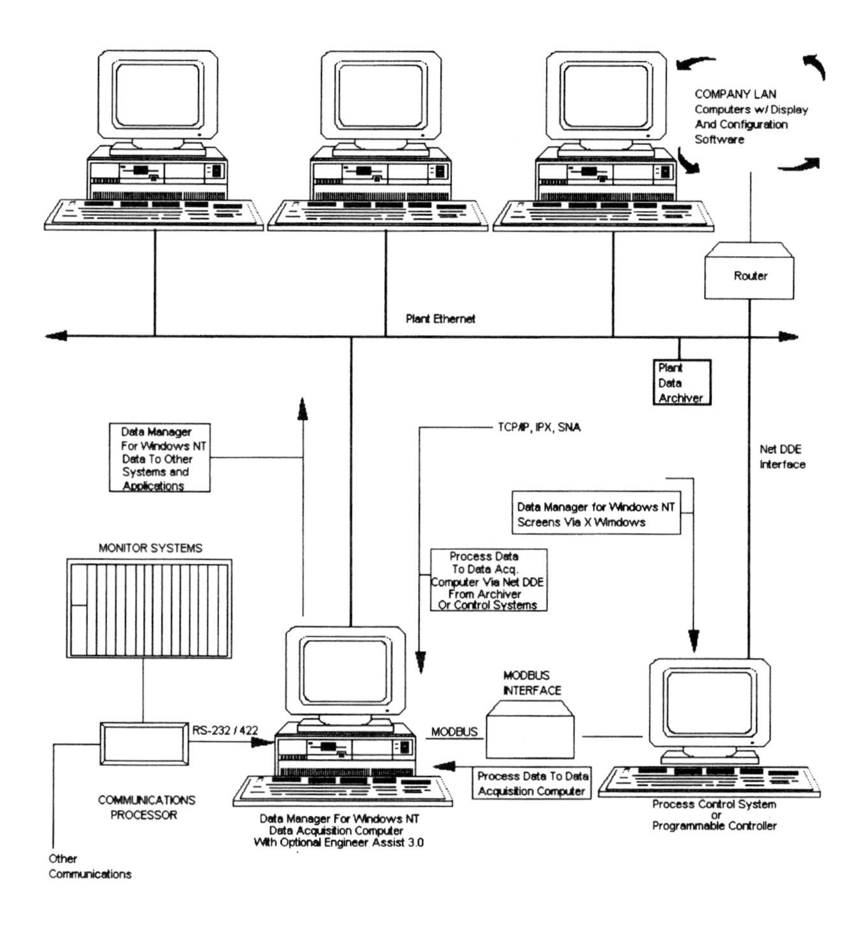

Figure 5 - On line continuous monitoring system for steady state and runup / rundown conditions, with process data access and resident knowledge based program.

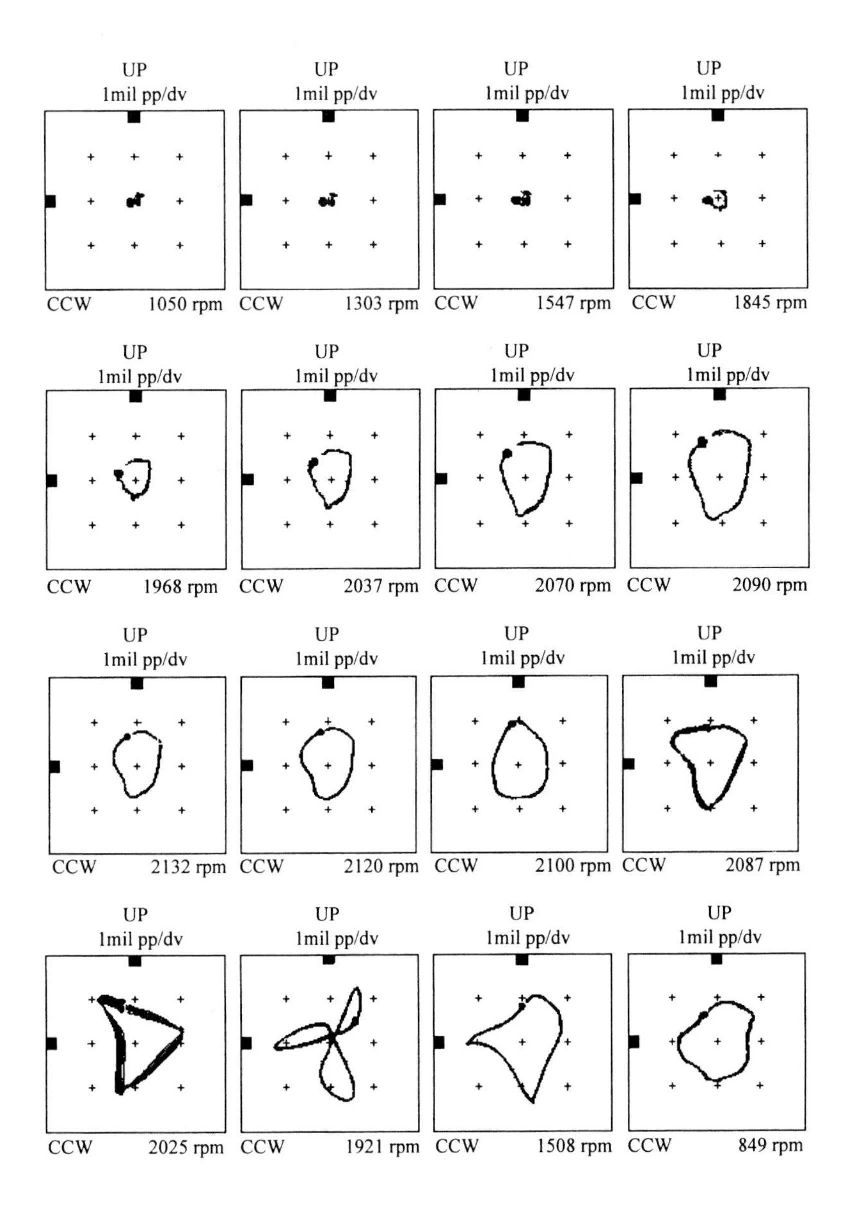

Figure 6 - Centrifugal compressor start up.

Showing the rotor response due to a radial rub which caused the shaft to bow and then impact. Note the thermally induced bow at 849 rpm on the rundown, compared with the unbowed rotor response at 1050 rpm during the start.

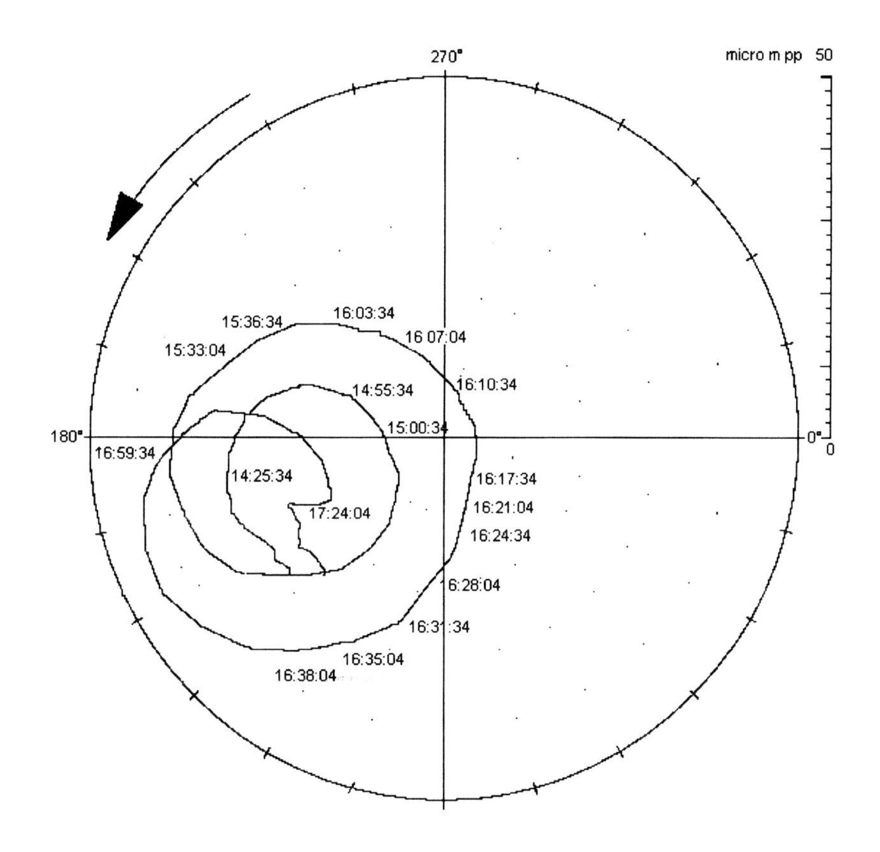

Figure 7 - Continual variation of 1X vibration vector produced by a radial rub which progressively modified the balance condition of the rotor.

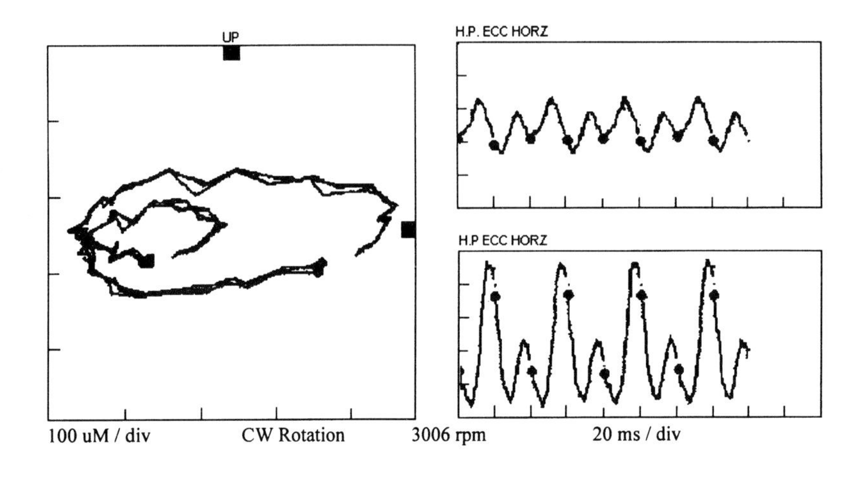

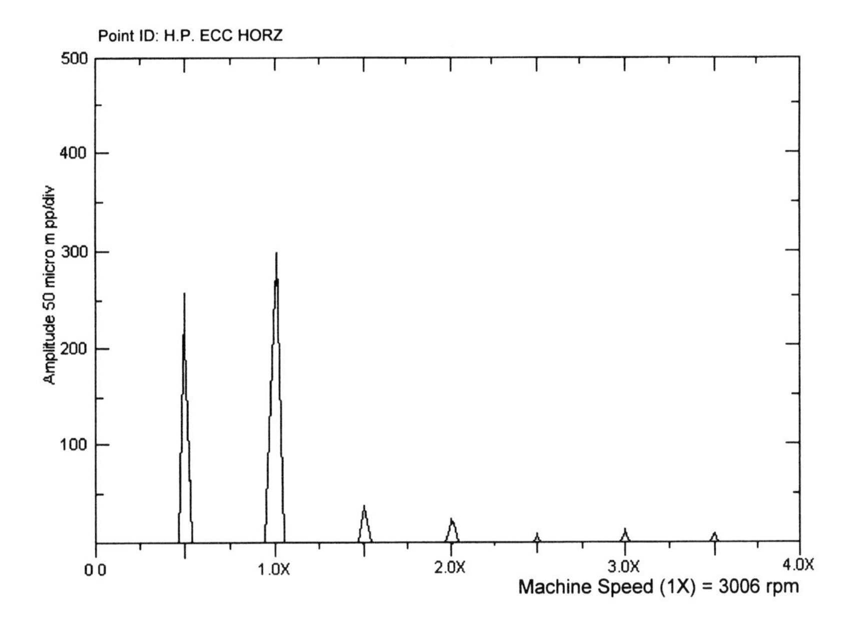

Figure 8 - Rub causing re-excitation of modified first balance resonance at exactly half running speed.

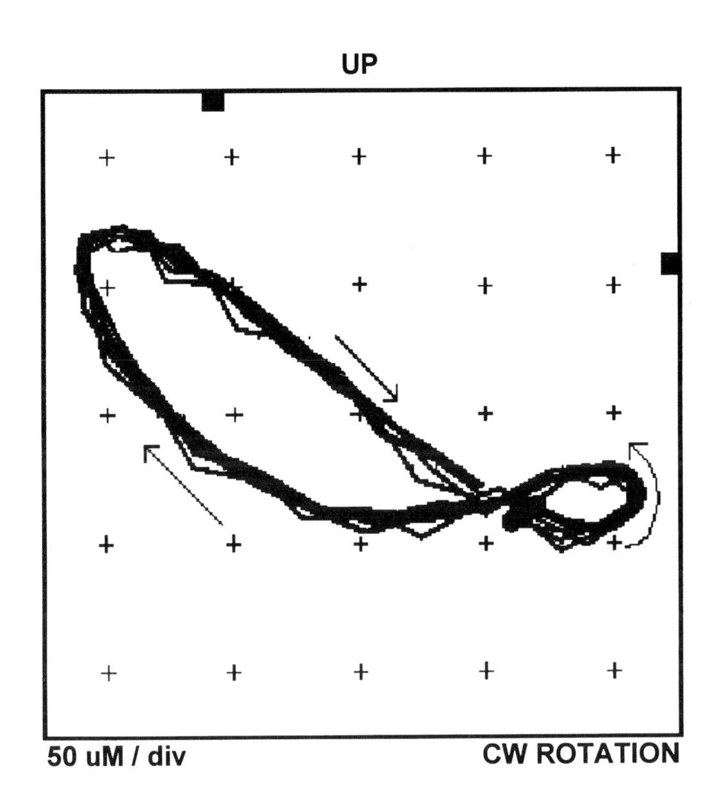

Figure 9 - Shaft orbit motion due to misalignment.

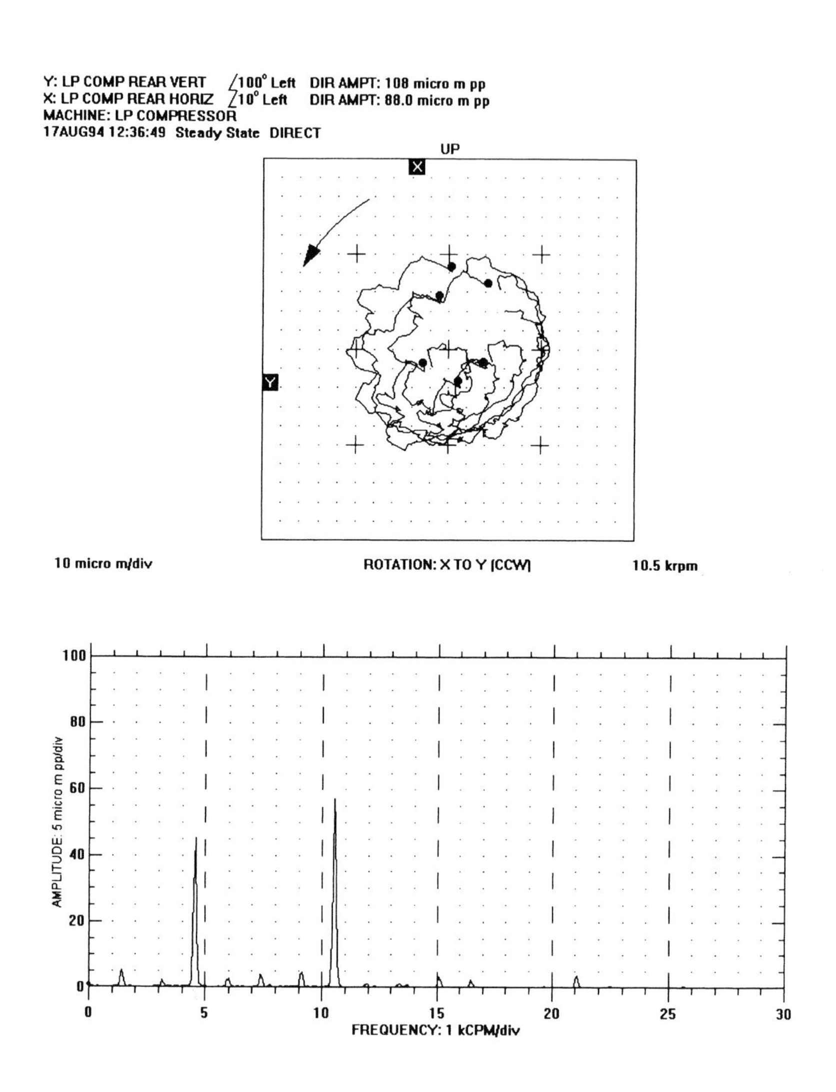

Figure 10 - Re-excitation of first balance resonance frequency due to fluid induced instability in a compressor seal.

S472/007/96

Computer-aided oil analysis as a predictive preventative maintenance tool

G E NEWELL
Oilab Lubrication Limited, Wolverhampton, UK

INTRODUCTION

Most power systems automotive, hydraulic, turbine, gear box and compressor operate with a circulatory lubricating fluid. This fluid is very similar to the blood that circulates around the human body. By analysing a sample of blood at any one time, would establish a condition of the efficiency of the circulatory system. Make assessments of your heart rate, establish the efficiency of the kidneys, (filter), and identify the level of contaminants (cholesterol). Apply the analagy to a lubricating oil, circulating around a system. That can also give you information about, first, is the oil performing? Is it doing its job? Do I need to change it? Is it contaminated? Also, what sort of wear is being generated and picked up by this lubricating oil fluid? Lubricant trend analysis programmes evaluate the condition of the fluid, the condition of the system, the environment in which it operates, and predicts, based on the data input, its future behaviour.

CONDITION MONITORING BY USED OIL TESTING

In a machine operating normally, certain oil test results tend to a constant rate of change. If this rate of change alters, then some change in operating condition has occurred. This may indicate a change in machine condition and possible incepient failure. One oil analysis result gives little information about machine condition.

Comparison of present results from the same (or closely similar) machines gives much more information. "To identify the abnormal we must first establish what is normal". The establishment of normality using a computer programme builds up a data bank of results relating to a particular piece or group of machinery and leads to the development of a trend analysis programme.

This will predict into the future the trends which are being established through oil testing, both on the physical condition of the lubricant the generation of wear debris.

Tests that establish the physical condition of the lubricant are as follows:-

Viscosity
Insolubles
Acidity (TAN)
Basicity (TBN)

Also contaminants in the lubricant such as water, fuel and in gas fuelled engines, hydrogen sulphide, (an extremely corrosive gas). Other factors, such as the depletion of the additive in the oil and the wear element generated, both in the rate of production and size of particles that are being produced. Other elements of contamination, such as water, airborne dust, and abrasive materials (ie the air filter may not be working). In the salt water environment of the marine industry, the contamination of the lubricant with chlorides(extremely corrosive). How important is it to monitor the changes and what will happen if they are ignored?

A condition monitoring programme establishes a physical condition and highlights changes in that condition due to a variety of reasons. A maintenance operative puts the wrong oil in and the viscosity rises. There would be an increase in the fuel and power consumption. During cold start conditions, oil starvation would occur causing extremely high initial wear rates and component failure. If the viscosity is lowered, full film lubrication is not maintained. The ability of the oil to separate the interacting surfaces is lost, thereby generating high temperatures, high oxidation and high wear rates. Predictable mechanical failure onset.

If the acid number in “clean” circulatory oils (ie turbine, hydraulic, gear oils) is allowed to rise too high, then corrosion takes place effecting piston rings, cylinder bores and antifriction bearings. The weak acids tend to corrode yellow and white metal bearings. Total acid number is particularly relevant to “clean” circulatory oils, that class of lubricants, seperate from the detergent/dispersant petrol/diesel crankcase lubricants.

Total base number gives the oil the ability to neutralize acidic percursors infiltrating the crankcase from the sulphur compounds in the fuel. The higher the sulphur content of the fuel, normally the higher the total base number. If the total base number is allowed to deplete to a low level then it no longer has the ability to neutralize acids coming into the crankcase. Severe corrosion and wear can occur. With automotive crankcase lubricants, the remaining level of the detergent/dispersancy is also critical. For a lubricant to disperse and hold in suspension all the sludge, varnish and resin percursors which are generated by burning the fuel, it must have a reserve within the lubricant. These materials are solublised by the detergent/dispersant additive in the oil and are held in suspension. If an oil reaches a point when there is no longer a reserve of detergent/dispersancy, then the oil will tend to shed the material causing rings sticking and the piston cleanliness level drops rapidly.

Insolubles have little effect on wear, were the detergent/dispersancy is adequate. They can however cause increased viscosity and impair pumpability at startup. Severe contamination can cause oil gelling. Abrasive insolubles, such as airborne dust, debris introduced during maintenance or with the make up oil and wear debris can, and probably will cause severe rapid wear. In hydraulic systems particularly with the more sensitive, as in the aircraft industry, very minute traces of insoluble material or particulate can cause wear and malfunction. For obvious safety reasons the standards set for that particular industry are extremely high.

The presence of water in an oil reduces the efficiency of the additives, it accelerates oil oxidation, causes corrosion, reduces the service life of the lubricant and causes erosion damage in hydraulic systems.

The test programme establishes the physical condition of the lubricant and the presence of damaging contaminants. Corrective action may be indicated or a complete oil change may be required.

Used oil condemning limits have been established by ASTM, oil companies and other learned bodies. These limits vary for different systems. In a lube oil circulatory system, oil condition tends to equilibrium. Additive depletion and contaminant build up are compensated by the addition of fresh top up lubricant. With the aid of a computer data base software programme the results can be compared with these limits. Trending on a time base will also indicate when these limits will be reached.

To illustrate the effect of ignoring test results, compare, one, a test programme of a lube oil, monitored in a diesel crankcase, and two the relevant piston cleanliness and ring groove blocking results. The first illustration shows that at approximately 180 hours running time, indicated by the dotted line, viscosity, insolubles and TAN rise rapidly with a similar rapid decline in TBN. The second illustration indicates relevant ring groove filling and piston cleanliness rating. By superimposing this second illustration over the first, coincident with the rapid fall of TBN and rapid rise In viscosity, insolubles and TAN the rapid drop in piston cleanliness and equally rapid rise in ring groove filling is indicated. This latter condition leads to ring sticking, reduced engine efficiency, increased blow by, which accelerates the loss of efficiency and oil degradation.

It is usual with most monitoring programmes today that an outside laboratory is used to conduct an analysis service. This involves two major drawbacks. First of all, cost and second delay. It is not unusual that the time lapse between the actual taking of the sample and the receipt of the result can be anything up to 3 weeks. A major benefit of on-site analysis is that you get immediate results. Another benefit is that the laboratory assistant conducting the test programme is divorced from the environment under which the oil is in service. The lubricator responsible for lubrication achieves a closer understanding of the machines under his control. A side benefit is that in taking a sample, the lubricator ensures that the system has oil. Many a failure is caused through total lack of lubricant through badly managed procedures and maintenance control.

This diagram illustrates the efficiency of three methods of conducting wear production programmes. Depending on the system, spectroanalysis is only box efficient in sub micron particles. Practically, the ferrographic technique, although expensive, is ideal for the normal run of equipment (5–10*um* in hydraulic, automotive, gearbox, compressor and turbine). A simple magnetic plug inserted into the circulatory system is becoming more popular. The plug is withdrawn from the system, the debris transferred to a card and microscopically examined for nature and rate of production in size categories.

Development of lube oil test kits over the past ten years has made it possible for maintenance personnel to conduct their own programmes in-house. A relevant simple and comprehensive test programme supported by specially developed software indicates courses of corrective action to be taken based on the results from the test programme.

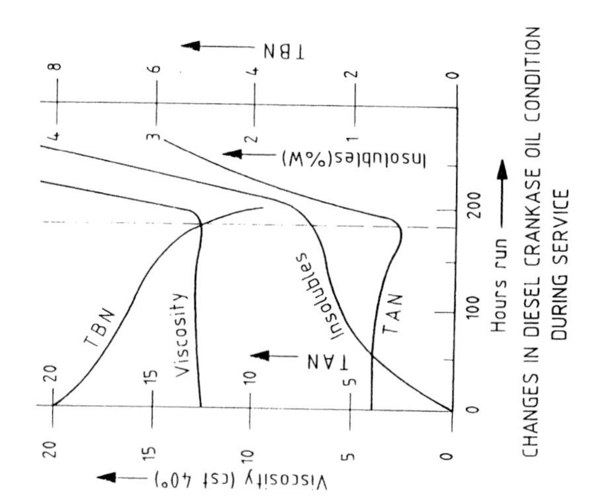

CHANGES IN DIESEL CRANKASE OIL CONDITION DURING SERVICE

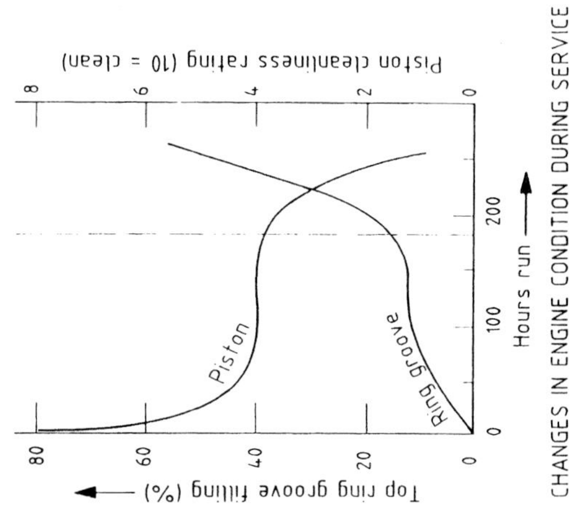

CHANGES IN ENGINE CONDITION DURING SERVICE

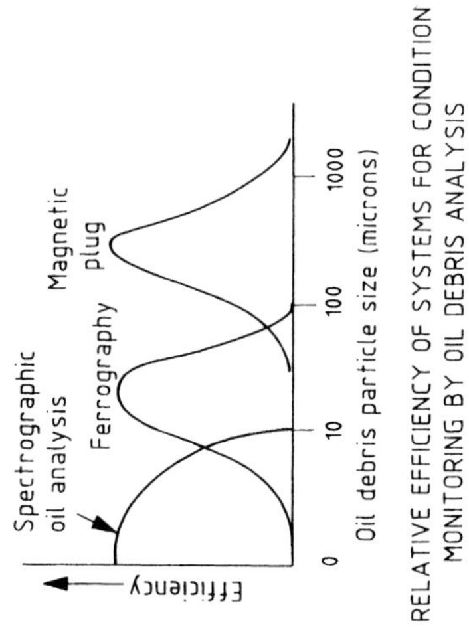

RELATIVE EFFICIENCY OF SYSTEMS FOR CONDITION MONITORING BY OIL DEBRIS ANALYSIS

The portable in house test programme has the following tests available:-

- Viscosity
- Water Content(0-1.2%)
- Trace Water Content(50–2000ppm)
- Presence Of Glycol
- Presence Of Salt Water
- Insolubles/Dispercancy
- Total Acid Number
- Total Base Number
- Pour Point
- Presence Of Abrasives
- Presence Of Hydrogen Sulphide
- ASTM Colour
- Wear Debris(Size/Nature/Volume)

In considering a relevant test programme for a gearbox (main coal conveyer drive) monitoring programme it would be structured as follows:-

- Viscosity
- Water Content(0-1.2%)
- Total Acid Number
- Wear Debris Production

Similarly for a hydraulic oil system (sub sea valve) activator: -

- Viscosity
- Trace Water
- Total Acid Number
- Particulate Evaluation

Wear production rates during run in are initially high before establishing, during normal running, a constant production rate both in size and volume of debris. Once this normal rate has been established then abnormal rates can be quickly detected. Incipient failure of equipment can be recognised by increased wear production rates and the increase in size of the wear particles. It is important to note that these rates differ for individual items of plant. One cannot place arbitrary figures on what is a critical wear production rate. These have to be established. "TO IDENTIFY THE ABNORMAL WE MUST FIRST KNOW WHAT IS NORMAL"

In support of the portable, on-site, lube oil analysis and wear monitoring equipment, OILAB have developed a computerised IBM compatible software programme. Physical test data obtained using OILAB equipment is measured against accepted standards and the illustration shows how the test figures are related to attention, warning, action and alarm limits. Based on succeeding test data input, a predictive trend is established in each test category. Using the trend predictive sequence, the computer indicates into the future when, based on the data input, these limits will be reached (see illustration). Using this information, sample taking intervals can be forecast and unnecessary sampling can be eliminated.

To summarise, I hope to have conveyed the importance of oil analysis in a preventative maintenance programme. Maintenance costs can be reduced by confident extension of oil change periods and critical wear production rates can be recognised enabling corrective action to be taken, prior to a catastrophic equipment failure and subsequent production stoppage.

In conclusion, the short & long term benefits that can be gained by oil analysis in a preventative maintenance programme are as below:-

Short Term

- Plant problem troubleshooting
- Projected time to critical failure
- Extension of oil drain intervals
- Plant evaluation prior to shutdown
- Overtime maintenance costs reduced by up to 30%

Long Term

- Predictable trend analysis identifying problems earlier in the failure cycle
- Extended service life of the lubricant
- Identifying poor maintenance practices
- Identifying stressed operating conditions
- Plant & equipment warranty support
- Overall maintenance costs reduced by up to 20%

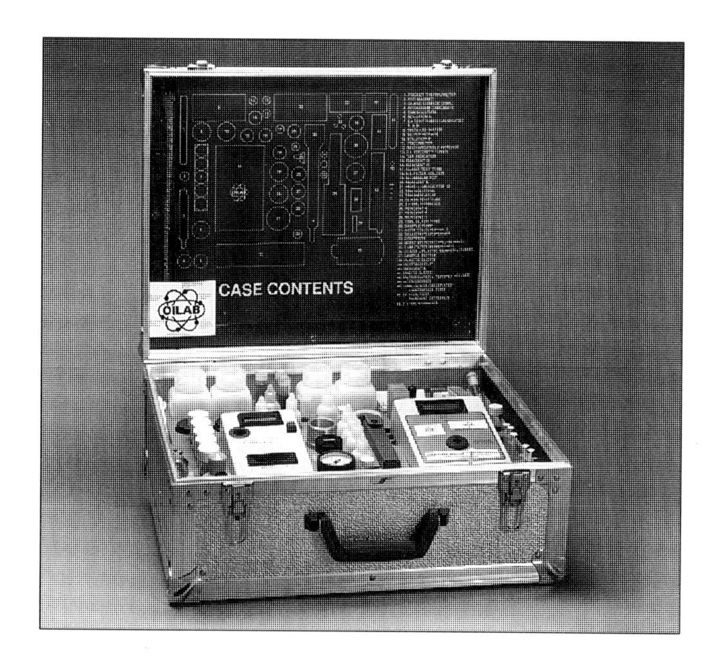

S472/008/96

Achieving true condition-based maintenance using artificial intelligence

J BIRD
Advanced Expert Systems Limited, Derby, UK

In recent years there has been an expansion in the application of condition monitoring techniques/systems. Significant sums of money are being invested in this field in an attempt to achieve the financial benefits of a true Condition Based Maintenance strategy. I would suggest that where these benefits are not being maximised the main reason is the quality of the interpretation of the condition data/information. This factor holds the key to the success of the whole condition monitoring strategy in that poor quality interpretations will rapidly lead to loss of confidence in the system/technique and will inevitably lead to the strategy being effectively abandoned.

In our experience organisations that have been able to attain high quality interpretations of their condition monitoring data/information are only then able to operate a true Condition Based Maintenance strategy thus maximising the financial returns.

High Quality Interpretations

The achievement of high quality interpretations is initially dependant upon reliable and timely sources of condition monitoring data. The interpretation of the health of complex machines such as engines, transmission units and turbines will probably involve the integration of several sources of data including maintenance and usage history. It is then necessary to evaluate the risk associated with a diagnosed fault and recommend the best means to minimise deterioration and restore machine health. All of these time consuming actions represent high usage of that most valuable resource, the engineer **(Figure 1.)**. Clearly there is a need for a tool that can relieve this workload from the engineer and help him to produce high quality consistent interpretations. We feel that recent developments in the field of Artificial Intelligence are able to fulfill this requirement.

Artificial Intelligence

Artificial Intelligence (AI) in general terms refers to techniques concerned with enabling computers to perform tasks commonly held to require human intelligence. Expert Systems is one of the widely used branches of AI and provides a programming capability for the replication of human expertise. Such programmes would consist of a hybrid of database programmes, statistical software and Expert System technology such that they are able to handle not only data and information but also expertise and reasoning **(Figure 2.)**.

If the Expert System can be designed to be generic in nature it can be utilised to integrate any combination of data/information from condition monitoring, performance measurement and operating parameters with maintenance and usage history to produce a diagnosis of the health of a machine. It can also provide a risk assessment for a particular fault and recommend actions to arrest deterioration and restore machine health. These features can be summarised as follows:

> **Data Validation**. This is an extremely important task as the quality of health assessment is directly linked to the quality of the data used. It includes checks for transcription, transposition, range etc.
>
> **Data Processing**. Often, data needs to be processed for it to be meaningful. Typically, this may mean filtering, smoothing, compressing, applying correction factors, comparing against a standard or passing through specialist statistics and engineering algorithms.
>
> **Interpretation**. The act of revealing symptoms concealed within the numerous sources of data.
>
> **Diagnosis**. Having revealed symptoms for a particular machine, diagnosis is required in order to determine which fault or faults may be inferred from the combination of symptoms present. In some cases the symptoms will represent a single fault, in others multiple faults may be represented.
>
> **Risk Evaluation**. Having determined possible faults, it is necessary to evaluate the risk associated with each so as to establish when action should be taken. This requires a review of primary and secondary consequences (prognosis) for each possible fault coupled with a forecast of the leadtime to failure remaining.
>
> **Countermeasures**. In some cases, it may be necessary to gather additional data, information, facts and symptoms in order to isolate a specific fault. Finally, appropriate countermeasures must be selected/formulated which will arrest the onset of failure and restore health.

The user and the Expert System work together as follows, **(Figure 3.)**:

- Condition data from a fleet of machines is entered into the programme.
- The System carries out the health assessment process for each machine separating out those developing problems from the healthy population and presenting clear and specific recommendations to the user.
- User takes decision and authorises a course of action for machines developing problems.

In this way the efficiency and effectiveness of machinery health management is greatly improved. **Figure 3.** illustrates the partnership between the user and the Expert System and shows the main flows of data and information. Wherever possible, the system provides recommendations from data fed in automatically, however in some cases it may request additional data, information, facts and symptoms from the user before arriving at a conclusion. Conversely, the user may request graphical displays of data, previous health assessments, maintenance history, decision logic etc. prior to accepting recommendations offered by the system. The Expert System can also form part of a "control loop" for it both detects deterioration and checks that deterioration has been arrested/health restored after actions have been implemented.

Industrial Applications

1. Aerator Gearboxes

A recent application of an Expert System was for the health assessment of 352 gearboxes transmitting power to the aerator impellers at the waste water treatment plant at Crossness Sewage Treatment Works. These gearboxes had been in service for 2 years when several failures occurred casting doubts on the suitability of the units for this duty. Within a short time the system was able to present an increasingly clear picture of the overall health of all the gearboxes on the site dispelling an earlier fear that the failure rate of the gearboxes would be high.

The advice presented by the system enabled engineers at Crossness STW to realise the following benefits:

Failure Prediction - Several gearboxes were diagnosed as being in an early stage of mechanical deterioration. Three such gearboxes were removed from service and the subsequent stripdown and examination confirmed the diagnoses by the system. All three gearboxes exhibited some gear wear but more significantly there was evidence of deterioration in one of the intermediate shaft bearings in each gearbox. The bearings and seals were changed and the units returned to service, previous failures had resulted in the need to replace components at an average cost of **£3000**.

Extended Oil Life - Substantial cost savings were achieved when the system indicated that the oil in 75% of the gearboxes was still suitable for continued use well beyond the manufacturers recommendation for oil change.

Inherent Gearbox Problems - The Expert System was instrumental in the confirmation of inherent problems within the gearboxes. The gearboxes are all of a similar design but are of 3 different sizes ie. 11Kw, 15Kw and 22Kw. It was able to pinpoint that the 11 K gearbox is more susceptible to water contamination and the 22 K gearbox is more likely to have dirt contamination due to an air gap on one sector of the joint between the gearbox and motor.

Overhauls - The system is providing feedback on the effectiveness of maintenance action and the quality of overhauls.

It is important to note that condition monitoring techniques in the form of oil analysis by spectroscopy and selected physical tests were already being used on the gearboxes prior to the introduction of the Expert System. The vast quantity of oil analysis data produced during this period was not only difficult to interpret but the workload resulted in the frequency of the oil sampling being extended well past the optimum level. Since the introduction of the Expert System the gearboxes are being sampled on a monthly basis and the health assessments produced are allowing the operators to adopt a cost effective Condition Based Maintenance strategy.

2. Dual Fuel Engines

Another application of an Expert System was to 4 large engines installed for power generation duties at Mogden Sewage Treatment Works and configured to run on a fuel mixture of 10% Diesel Oil and 90% Biogas. Each engine is able to generate 2.4Mw of power that is available to be exported to the regional electricity company. Originally designed and installed for a less arduous duty, these engines are now required to run at maximum load with a fuel that can not only vary in calorific value but can occasionally be heavily water laden and contain corrosive substances such as hydrogen sulphide.

The system was installed in July 1995 and immediately indicated an URGENT status on Engine No 2. **(Figure 4.)** with 2 distinct faults (cooling leaks and excess blowby). A major scheduled overhaul of the engine was brought forward and the subsequent examination of the engine components confirmed the diagnoses.

> **Coolant leaks** - This was the most pressing problem and caused the URGENT flag to be raised in the condition indicator. On stripdown there was evidence that at least one of the exhaust valve water jackets had developed a leak and four of the rolled tubes that seal the cylinder head where the cylinder head bolts pass through had also been leaking.
>
> **Excess Blowby** - There were two indicators to support this diagnosis ie. scuffing had occurred inside several of the cylinder liners and some corrosion/wear was visible on the small end bushes. These bushes are copper based and this corrosion is due to H_2S in the fuel bypassing the piston during blowby and contaminating the lube oil.

The repair of the engine was carried out at minimum cost due to the timing of the overhaul. It was generally accepted that if the overhaul had been delayed awaiting the recommended service interval hours then greater damage would have occurred resulting in an increase in the cost of the overhaul. This saving was estimated to be in the region of £100,000 and was the result of recommendations made 6 days after the commissioning date of the Expert System. It can be justifiably claimed that this case represents the first major step towards Condition Based Maintenance.

On Engine No 3. the health assessment by the system **(Figure 5.)** confirms this engine to be reliable with no signs of problems. Although this engine is due for a major overhaul, this is to be deferred until recommended by the system. This is another good example of how an organisation is practising Condition Based Maintenance.

Again oil analysis techniques of spectroscopy and selected physical tests were applied to these engines prior to the introduction of Expert Systems but they were not able to prevent the catastrophic failure of one of the engines in 1994. Thames Water, the sponsor of these two particular projects, are sufficiently convinced as to the benefits of the use of Expert Systems that a proposal has been put forward to install the system on all their large engines (150), gas turbines and the pump units on the London Ring Main system. Thames Water have included the System in their register of 'Best Practice' procedures that is circulated throughout the organisation.

Conclusions

The common factor in these applications is how a previously unsuccessful condition monitoring regime was transformed by the introduction of an Expert System that was able to increase the level of confidence in the interpretations and recommendations. This transformation was instrumental in all the organisations achieving the financial benefits of a true Condition Based Maintenance strategy.

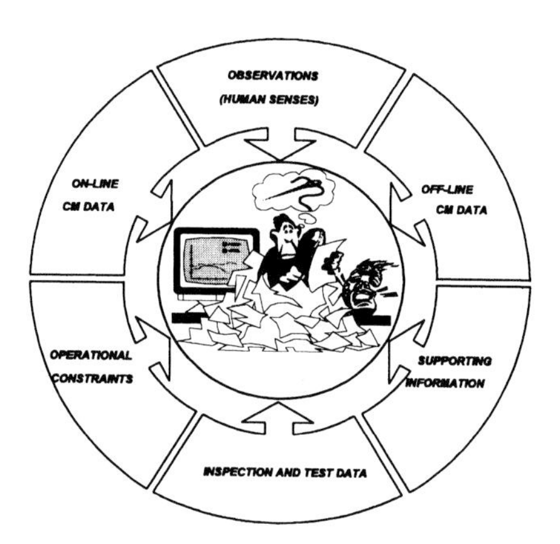

Figure 1. Current Situation

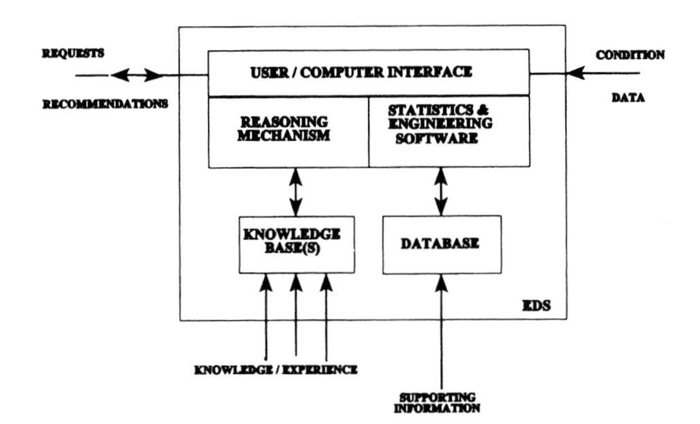

Figure 2. PHOCUS Expert Decision Support System

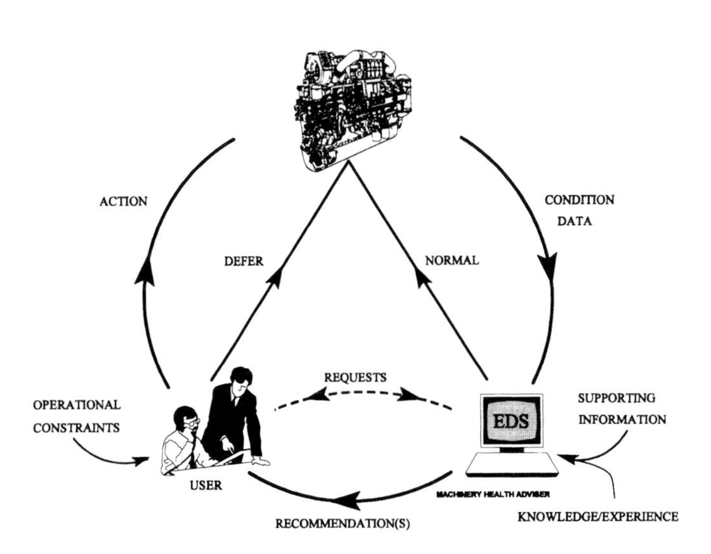

Figure 3.

RECOMMENDATIONS

Unit: Power House	Model: MIRRL_KP6_MK3	Last OC:
Comp: EA2	Lube: BP_7472_USED	Last ME:
Name: Engine No 2.	Coolant: WES 635	Homeshop: MOG

Sample ID	DDMMYY	S	Condition Indicators	Action
MOG_68	240795	U	URGENT : Inv Immediately Possible Blowby Possible Si from Dirt/Ingress Na Exceeds Lube Change Limit HLR Si (36 ppm) Lim 30 Si is Stable HLR Na (109 ppm) Lim 28.2 Na is Stable HLR Insolubles (0.51%) Lim 0.5	Chg Lube & Filter Ck injectors Ck Excess Blowby Ck Air Filters Ck intakes/ sand Ck coolant Leaks Ck Gas Settings
MOG_67	070795	C	URGENT: Inv Immediately Possible Blowby Possible Si from Dirt/Ingress Na Exceeds Lube Change Limit HLA Si (29ppm) Lim 20 Si is Stable HLR Na (108 ppm) Lim 28.2 Na is Increasing Rapidly HLA Insolubles (0.42%) Lim 0.21	Chg Lube & Filter Ck Injectors Ck Excess Blowby Ck Air Filters Ck Intakes/Sand Ck Coolant Leaks Ck Gas Settings
MOG_51	130695	C	URGENT : Inv Immediately Na Exceeds Lube Change Limit HLR Na (94 ppm) Lim 28.2 Na is Increasing Rapidly HLA Insolubles (0.47%) Lim 0.21	Chg Lub & Filter Ck Injectors Ck Excess Blowby Ck Coolant Leaks Ck Gas Setting
MOG_50	050695	C	URGENT : Inv Immediately Possible Si from Dirt/Ingress Na Exceeds Lube Change Limit HLA Si (29 ppm) Lim 28.3 Si is Stable HLR Na (51 ppm) Lim 28.2 Na is Increasing Rapidly HLA Insolubles (0.4%) Lim 0.21	Chg Lub & Filter Ck Injectors Ck Excess Blowby Ck Air Filters Ck Intakes/Sand Ck Coolant Leaks Ck Gas Setting
MOG_49	220595	A	CAUTION/ALERT : Monitor Engine Closely HLA Insolubles (0.4%) Lim 0.21	Ck Injectors Ck Gas Setting Ck Excess Blowby

Figure 4.

RECOMMENDATIONS

Unit: Power House	Model: MIRRL_KP6_MK3	Last OC:
Comp: EA3	Lube: BP_7472_USED	Last ME:
Name: Engine No 3.	Coolant: WES 635	Homeshop: MOG

Sample ID	DDMMYY	S	Condition Indicators	Action
MOG_90	021095	N	Normal Condition	
MOG_89	200995	N	Normal Condition	
MOG_88	080995	N	Normal Condition	
MOG_87	240895	N	Normal Condition	
MOG_86	040895	N	Normal Condition	
MOG_73	280795	N	Normal Condition	
MOG_72	240795	N	Normal Condition	
MOG_71	070795	N	Normal Condition	
MOG_70	130695	N	Normal Condition	
MOG_60	050695	N	Normal Condition	
MOG_59	220595	N	Normal Condition	
MOG_58	050595	N	Normal Condition	
MOG_57	010595	C	URGENT : Inv Immediately HLR Na (29 ppm) Lim 28.2 Na is Increasing Rapidly	Ck Coolant Leak
MOG_56	240495	A	CAUTION/ALERT : Monitor Engine Closely HLA Viscosity (119) Lim 118	Ck Fuel Timing Ck Gas Setting
MOG_55	180495	N	Normal Condition	

Figure 5.

Authors' Index